Sustainable Smart Cities

Priyanka Mishra • Ghanshyam Singh

Sustainable Smart Cities

Enabling Technologies, Energy Trends
and Potential Applications

Priyanka Mishra
University of Johannesburg
Johannesburg, South Africa

Ghanshyam Singh
University of Johannesburg
Johannesburg, South Africa

ISBN 978-3-031-33353-8   ISBN 978-3-031-33354-5 (eBook)
https://doi.org/10.1007/978-3-031-33354-5

© The Editor(s) (if applicable) and The Author(s), under exclusive license to Springer Nature Switzerland AG 2023
This work is subject to copyright. All rights are solely and exclusively licensed by the Publisher, whether the whole or part of the material is concerned, specifically the rights of translation, reprinting, reuse of illustrations, recitation, broadcasting, reproduction on microfilms or in any other physical way, and transmission or information storage and retrieval, electronic adaptation, computer software, or by similar or dissimilar methodology now known or hereafter developed.
The use of general descriptive names, registered names, trademarks, service marks, etc. in this publication does not imply, even in the absence of a specific statement, that such names are exempt from the relevant protective laws and regulations and therefore free for general use.
The publisher, the authors, and the editors are safe to assume that the advice and information in this book are believed to be true and accurate at the date of publication. Neither the publisher nor the authors or the editors give a warranty, expressed or implied, with respect to the material contained herein or for any errors or omissions that may have been made. The publisher remains neutral with regard to jurisdictional claims in published maps and institutional affiliations.

This Springer imprint is published by the registered company Springer Nature Switzerland AG
The registered company address is: Gewerbestrasse 11, 6330 Cham, Switzerland

Preface

Currently, the world is becoming increasingly urbanized, and the projected forecast reveals that more than 60% of the population will reside in the city by 2030. The cities and metropolitan areas are powerhouses of economic growth—contributing about 60% of global gross domestic product (GDP). However, they also account for about 70% of global carbon emissions and over 60% of resource use. Thus, rapid urbanization results in a growing number of slum dwellers, inadequate and overburdened infrastructure and services, worsening air pollution, and unplanned urban extension. Urbanization can threaten the quality of life, but cities also provide the testbeds needed to come up with related solutions. It will only be through the collective efforts of governments, private sector, and public—by harnessing transformative technologies—that we will be able to rebuild in ways that realize the true potential of cities. Smart City is a global movement in the use of technology for innovation, community development, and enhancing the quality of life of its inhabitants with a focus on sustainability. The well-integrated use of modern technology promotes community growth and development. Its potential advantages include social growth by providing several avenues for innovation, new business creation, and idea generation. In addition, it reduces the government's public spending on correcting public service management and provides efficient high-quality services because of effective process management. However, starting with a broad definition, the use of Information and Communication Technologies (ICTs) in the process of creating a more sustainable city, as well as the availability and quality of knowledge communication and social infrastructure, is central to the Smart City concept and distinguishes it from "sustainable cities." Smart governance, smart transportation, smart energy, smart healthcare, smart factory, smart technology, smart security, smart agriculture, and smart education are the key dimensions that smart cities can be identified along. Smart cities have emerged as a potential solution to the environmental issues that have arisen because of increased urbanization. These are very demanding and necessary for the long-term future. A smart sustainable city is an innovative city that uses ICTs and other means to improve quality of life, the efficiency of urban operations and services, and competitiveness, while also meeting the economic, social, environmental, and cultural needs of current and future generations. In smart cities, artificial intelligence (AI) helps to make urbanization smarter and gives inhabitants more convenient and sustainable solutions. Every city in the world requires healthcare as a basic essential service. Smart healthcare employs technology such as the Internet of Things (IoT), AI, wearable devices and sensors, Big Data, Business Intelligence, and others to intelligently adapt to the healthcare ecosystem's needs. IoT technology allows for real-time data collection via medical sensors, as well as data analysis for accurate medical analysis. Wearable gadgets and virtual assistants have enabled patients to track and monitor their health issues on their own. The use of digital technologies improves communication across different departments in a healthcare organization, resulting in faster services. This book delivers a recent rationalized theoretical model, which can be used for practical purposes of future city. Further, it also highlights ongoing trends, revealing some fresh research data results that might be interesting for researchers when designing smart cities. Currently, we are living in an information era society and knowledge has been explosively accumulating; therefore, it is impossible to embody all new developments in one book. However, in this book, we have tried to cover most representative achievements with emphasis on thoughts, philosophy, and methodology of sustainable smart cities.

At present, a dedicated book to discuss about the importance, developments, needs, and sustainability in smart cities. Its significant roles in governance, transportation, energy, healthcare, factories, technologies, securities, agriculture, and education. The impact of enabling technologies on smart cities and how they enhance citizens lives, information sharing with the public, and come up with a finer attribute of citizen welfare and government services. Further, the proposed framework and architecture of IoT-based sustainable smart city with sixth-generation (6G) system is the unique feature of this book. The use of AI in many zones of cities to increase the system's performance and efficiency is depicted quite well. With the detailed discussion of energy management in smart cities, we have also presented how Internet of Vehicles (IoV) uses wireless communication and sensing technology to establish a network of information exchange between vehicles, infrastructure, and the environment. Moreover, the upgradation of healthcare systems, Internet of Medical Things (IoMT) as well as Unmanned Aerial Vehicles (UAVs) for IoT-based sustainable smart cities is also presented. This book explores societal, economic, and practical reforms that would promote smart cities based on a variety of worldwide case studies. Eventually, academicians/researchers will profit because it will cover all aspects of sustainable smart cities in one book.

- Overview and discussion of the importance of sustainable smart cities, characteristics, requirements, and distinct architectural design of Sustainable Smart Cities as compared to that of the existing conventional cities.
- Provides an extensive overview and state-of-the-art enabling technology that can be put into practical use in order to assist cities prosper.
- Explored 6G-IoT-based framework and architecture for sustainable smart cities with its potentialities and open research challenges.

- Presented the role of artificial intelligence for potential key components of sustainable smart cities such as intelligent transport systems, smart and connected healthcare, smart energy systems, smart industries, smart education, and digital farming.
- Explored the significance of connected and cognitive technologies for improved monitoring and real-time analytics of the key components of sustainable smart cities.
- Exploited the role of IoV and UAVs with their vast range of applications and significant contribution in a variety of domains to the development of the sustainable smart city.
- Addresses recent advancements, along with relevant prospects on opportunities and challenges with potential research direction in the development of sustainable smart cities.

The topics covered in the book are concerned with sustainable smart cities and their development for better life of inhabitants and industries. In this way, it will attract various stages of readers as given below:

1. This book will assist researchers working in academia and industry in conceptualizing the development of future sustainable smart cities.
2. It encourages researchers in academia and industry to learn more in this evolving field of sustainable smart cities.
3. This book brings together a comprehensive collection of significant research works on establishing sustainable smart city from the perspectives of IoT and AI.
4. Integration of enabling technologies for the development of sustainable smart cities based on next-generation communication systems.

This book is intended for senior undergraduate and graduate students working in electronics/electrical and communications engineering; however, it can also be used as a reference book for engineering professional and scientists working in academia and industry. In this book, we have tried to cover most representative achievements, with emphasis on thoughts, philosophy, and methodology for sustainable smart city accomplishment. Indeed, existing knowledge is undoubtedly important. However, in this book, we have provided our thoughts and methodology which helps in creating new knowledge for the future sustainable smart city.

This book comprises 10 chapters. Chapter 1 provides a comprehensive overview of the state-of-the-art smart city which is a digitally connected urban environment that uses data, technology, and analytics to improve the quality of life for its inhabitants. It utilizes sensors and other digital technologies to enable citizens to access services quickly and efficiently. Smart cities also prioritize sustainable development and innovation. The objectives of a smart cities are to use technology and data to improve the efficiency of services such as energy, transportation, and waste management, as well as to improve the quality of life of citizens. Smart cities also aim to reduce carbon emissions and promote sustainability. It is built on the idea of utilizing technology to enhance the lives of citizens and create a more efficient and sustainable urban environment. They are enabled by the use of a variety of data-driven technologies such as the IoT, AI, and big data. This data can be used to develop more efficient ways of managing the city's resources and improve services for citizens. The purpose of this chapter is to examine the importance and need of sustainable smart cities with examples from around the world that offer examples. Further, a detailed aspect of smart city with its requirements, potential components, and open research challenges with opportunities are discussed.

Chapter 2 discusses how the growth of data traffic and the interconnection of digital devices have made the internet an essential resource in society, with a potential for increased demand. To develop a system that securely connects the internet to real-world space would aid in the advancement of a human-centered society that balances economic progress with the resolution of social issues. This chapter provides a detailed overview of Society 5.0 with its requirements, architecture, and components. Society 5.0 has been extensively explored in detail, its relation to Industry 4.0/5.0, as well as the role of Society 5.0 in the Sustainable Development Goals of the United Nations. Several emerging communication and computing technologies such as 5G/6G-IoT, edge computing/cloud computing/fog computing, Internet of Everything (IoE), blockchain, and beyond networks have also been explored to fulfil the demands of Society 5.0. The potential application of super smart cities (Society 5.0) with some practical experience of inhabitants is thoroughly discussed. Finally, we highlight several open research challenges with opportunities.

Chapter 3, aims to achieve Sustainable Development Goal (SDG) 11, which focuses on making cities inclusive, safe, resilient, and sustainable. Smart cities are expected to play a crucial role in improving the efficiency and effectiveness of urban management, encompassing areas such as public services, public security, and environmental protection. In the development of smart cities, enabling technologies has been identified as a key enabler. Globally, cities need intelligent systems to cope with limited space and resources. As a result, smart cities emerged mainly as a result of highly innovative ICT industries and markets, and additionally, they have started to use novel solutions taking advantage of the IoT, big data, and cloud computing technologies to establish a profound connection between each component and layer of a city. In this chapter, we have discussed several key technologies including IoT, IoE, wireless sensor network, computing technologies, 5G, 6G, cloud computing, edge computing, fog computing, quantum communication, artificial intelligence, machine learning, deep learning, big data/data analytics, blockchain, and cyber physical systems (CPS) that contribute to the creation of a smart city. It also offers advanced solutions for today's cities.

In Chapter 4, we explore how the IoT serves as the backbone of smart cities, enabling various aspects of smart cities to become more accessible and applicable. The chapter focuses on the concept of sustainable smart cities and examines the different applications, benefits, and advantages associated with the integration of IoT technologies. It also explores the potential of smart cities to drive technological advancements in the future, while addressing open issues, and research challenges associated with their implementation.

Preface ix

In Chapter 5, we have explored the vision of sixth-generation (6G) Internet of Things-enabled communication networks for sustainable smart cities with its intelligent framework. In this chapter, we proposed framework consists of real-time sensing, computing, and communication systems. Further, we have emphasized the architecture and requirements of the 6G-IoT network in relation to the increasing demand for the establishment of a sustainable smart city. In addition, we describe emerging technologies for 6G in terms of artificial intelligence/machine learning, spectrum bands, sensing networks, extreme connectivity, new network architectures, security, and trust that can enable the development of 6G-IoT network architectures to ensure the quality of service (QoS)/quality of experience (QoE) that is large enough to support a massive number of devices. Moreover, we have highlighted potential challenges and research directions for future 6G-IoT wireless communication research for the sustainable smart cities.

In Chapter 6, an application of AI in the sustainable smart city operations is presented. The AI includes expert systems, natural language processing, speech recognition, and machine vision. The use of information and communication technologies boosts the economy, improves the lives of citizens, and assists the government in its duties. The AI can play a significant role in making cities safer and easier to live in. It gives people more control over their own homes, monitors traffic, and manages waste. A smart city's sustainable development is investigated in this chapter through the application of AI. These applications could be employed in smart transport, smart healthcare, smart homes/living, smart industries, smart energy, smart agriculture, smart governance, and smart education, Further, existing daily lives and social media examples of AI in smart cities are discussed in detail.

Chapter 7 concerns with energy conservation systems for managing energy in smart cities as well as buildings. With the IoT, smart cities can benefit from a number of smart and extensive applications. As IoT devices are becoming more energy efficient, their overall energy consumption is increasing due to their proliferation and the increasing number of connected devices. Therefore, it is imperative to implement solutions that effectively manage energy usage in order to address this growing demand. In the smart cities, energy management is often approached from an interpretivist perspective. In this regard, smart city solutions must take into account how energy must be used while also dealing with the issues related to it. Throughout this chapter, we provide a view of energy management and issues facing smart cities. Furthermore, we presented a comprehensive classification of energy management for Internet of Energy (IoE)-based smart cities. Additionally, we discussed energy harvesting in sustainable smart cities, which can extend the life of low-power devices, as well as the challenges that come with it.

Chapter 8, emphasis on how transport system plays a vital role in large urban areas and in modern life. It enable the mobility of not only vehicles, but also people residing in cities. The issue of mobility in urban environments is vitally relevant to them. On the IoT, vehicles become intelligent moving nodes or objects within the sensor network, combining the mobile internet and the Internet of Vehicles (IoV). With an emphasis on deployment in smart cities, this chapter examines the current and emerging IoV paradigms and communication models. The main purpose is to assess the ability of the IoV to serve the needs of smart cities in terms of data collection, processing, and storage on a large scale.

Chapter 9 discusses about the concept of smart healthcare system in sustainable smart cities. It shows how traditional medical system is transformed into a smart healthcare system in a broader sense by using cutting-edge information technologies, like IoT, big data, cloud computing, and AI. Next, we discuss smart healthcare issues and its possible solutions. In conclusion, we look forward to the future of smart healthcare and evaluate its prospects which results in more efficient, convenient and personalized healthcare.

Finally, an unmanned aircraft system (UAS)/unmanned ariel vehicle (UAV) is presented in chapter 10. Its application is used to monitor and regulate urban development, transportation, monitor air quality, inspect infrastructure, and provide emergency services, among other things. This can also be used to deliver goods, reducing traffic and emissions from traditional delivery vehicles. In addition, it provides real-time data to service provider and governing authorities of smart cities for better decision-making and resource allocation. This could help create more sustainable, efficient, and equitable smart cities. Further, we discuss how UAVs can help to solve problems and build a smart city sustainable. Finally, we evaluate UAVs' prospects and look forward to its future research scopes.

In summary, the book provides a unified view of the state-of-the-art sustainable smart city, which should be accessible to the readers with basic knowledge of emerging technology and their role to accomplish sustainable smart city. The readers may find the extensive references provided in each chapter to be highly valuable and beneficial. The authors have performed a good job by providing a concise summary of all the chapters at the preface of the book. I would strongly recommend the book to senior undergraduate/graduate students, researchers, and engineers working or intending to work in the area of sustainable smart city and emerging technologies required to establish sustainable smart city. Although numerous journal/conference publications, tutorials, and books on the sustainable smart city have been published in the last few years, the vast majority of them focus on the various components of the smart city. However, this book distinguishes itself from the existing prosperous literature on sustainable smart city and its enabling technologies.

The authors are greatly indebted to numerous colleagues for the valuable suggestions during the entire period of manuscript preparation. Dr. Priyanka Mishra is thankful to her mentor Prof. (Dr.) Ghanshyam Singh for providing an opportunity and motivation to explore the potential and work in the field of sustainable smart city. She also conveys greatest gratitude to her parents whose tremendous blessings and endless sacrifice always encourages during the writing of the manuscript as well as carrier growth. They have shown immense patience and supported in every possible way. Words alone can never express the gratitude to them. We would also like to thank Springer, in particular Charles B. Glaser, Brian P. Halm, and Dhivya Savariraj for their helpful guidance and encouragement during the creation of this book. The authors would not justify their work without showing the gratitude to their family members who have always been the source of strength to tirelessly work to accomplish the assignment. Last but not least, we thank the ultimate source of energy of every particle in the universe, the Almighty, for giving the enough energy and strength to complete the work. All praise and gratitude belong to him.

Johannesburg, South Africa

Priyanka Mishra
Ghanshyam Singh

Contents

1	**Introduction: Importance of Sustainable Smart City**	**1**
1.1	Introduction: Smart City and Its Demands	1
1.2	Sustainability in Smart City	6
1.3	Definition of Smart and Sustainable Smart City	8
1.4	Driving and Guiding Principles to Create Sustainable Smart Cities	11
	1.4.1 Smart for Everyone and Everywhere	12
	1.4.2 Technology as an Enabler	12
	1.4.3 The Local Context	13
	1.4.4 Needs of the City and Community	13
	1.4.5 Innovation, Alliances, and Participation	13
	1.4.6 Sustainable, Secure, and Robust	14
1.5	Potential Components of Sustainable Smart City	14
	1.5.1 Smart Governance	14
	1.5.2 Smart Transportation	15
	1.5.3 Smart Energy	16
	1.5.4 Smart Healthcare	16
	1.5.5 Smart Industry	17
	1.5.6 Smart Technology	17
	1.5.7 Smart Buildings and Infrastructure	17
	1.5.8 Smart Agriculture	18
	1.5.9 Smart Education	18
1.6	Requirements of Smart City	19
1.7	Key Performance Indicators of Sustainable Smart Cities	20
	1.7.1 Economic Sustainability	20
	1.7.2 Ecological Sustainability	22
	1.7.3 Social Sustainability	24
1.8	Smart Cities Around the World	25
1.9	Open Research Challenges and Research Direction	29
1.10	Organization of the Book	30
1.11	Conclusion	33
	References	33

xiii

2	**Sustainable Smart City to Society 5.0**	39
	2.1 Introduction	39
	2.2 Sustainable Smart Cities Toward the Industrial Revolution	43
	2.2.1 Industry 4.0/5.0	43
	2.2.2 Society 5.0	44
	2.3 Requirements of Society 5.0	47
	2.3.1 Anxious Feelings of Inhabitants Regarding the Realization of Society 5.0	47
	2.4 United Nations Sustainable Development Goals (SDGs) for Society 5.0	49
	2.5 Open Research Challenges and Research Directions	51
	2.5.1 Healthcare	52
	2.5.2 Mobility	53
	2.5.3 Infrastructure	53
	2.5.4 Fintech	53
	2.6 Summary	54
	References	54
3	**Enabling Technologies for Sustainable Smart City**	59
	3.1 Introduction	59
	3.2 Recent Advances in Communication Technologies	64
	3.2.1 5G/6G	64
	3.2.2 Quantum Communication	65
	3.3 Internet of Things (IoT)/Internet of Everything (IoE)	65
	3.4 Artificial Intelligence (AI)/Machine Learning (ML)/Deep Learning (DL)	66
	3.5 Big Data/Data Analytics	67
	3.6 Blockchain	68
	3.7 Cyber-Physical Systems (CPS)	68
	3.8 Wireless Sensor Networks (WSN)	68
	3.9 Open Research Challenges and Research Directions	69
	3.10 Summary	69
	References	70
4	**Internet of Things for Sustainable Smart City**	75
	4.1 Introduction	75
	4.2 IoT Architecture for Sustainable Smart Cities	78
	4.2.1 Application Layer	79
	4.2.2 Transmission Layer	79
	4.2.3 Data Management Layer	80
	4.2.4 Sensing Layer	80
	4.3 Computing Technologies in IoT Environment	81
	4.3.1 Cloud Computing	82
	4.3.2 Edge Computing	82
	4.3.3 Fog Computing	83
	4.3.4 Quantum Computing	84

	4.4	IoT Applications in Sustainable Smart Cities	84
	4.5	Open Research Challenges and Research Directions	90
	4.6	Summary	92
	References		93
5	**6G-IoT Framework for Sustainable Smart City: Vision and Challenges**		**97**
	5.1	Introduction	97
		5.1.1 Evolution of Generations: 1G to 6G	99
	5.2	Vision and Requirements	102
	5.3	6G-IoT Framework for Sustainable Smart City	104
		5.3.1 Sensing Layer	104
		5.3.2 Network Layer	105
		5.3.3 Communication Layer	105
		5.3.4 Data Processing Layer	107
		5.3.5 Application Layer	107
		5.3.6 Business Layer	107
	5.4	6G-IoT Enabling Technologies	108
		5.4.1 Artificial Intelligence and Machine Learning	108
		5.4.2 New Spectrum Technologies	109
		5.4.3 A Network That Can Sense	109
		5.4.4 Extreme Connectivity	109
		5.4.5 New Network Architectures	110
		5.4.6 Security and Trust	110
	5.5	6G-IoT Use Cases	110
		5.5.1 Smart City/Home	111
		5.5.2 Smart Healthcare	112
		5.5.3 Satellite Terrestrial Network	112
		5.5.4 Marine and Underwater Communication	113
		5.5.5 Extended Reality (XR)	113
	5.6	Open Research Challenges and Research Directions	113
	5.7	Summary	114
	References		115
6	**Artificial Intelligence for Sustainable Smart Cities**		**119**
	6.1	Introduction	119
	6.2	Machine Learning/Deep Learning for Sustainable Smart Cities	121
		6.2.1 Machine Learning for Smart Cities	123
		6.2.2 Deep Learning for Smart Cities	124
	6.3	Recent Applications of Artificial Intelligence for Sustainable Smart Cities	124
		6.3.1 Artificial Intelligence for Smart Transport	124
		6.3.2 Artificial Intelligence for Smart Industries	126
		6.3.3 Artificial Intelligence for Smart Home/Smart Living	126
		6.3.4 Artificial Intelligence for Smart Healthcare	127
		6.3.5 Artificial Intelligence for Smart Energy	127

		6.3.6	Artificial Intelligence for Smart Agriculture	128
		6.3.7	Artificial Intelligence for Smart Governance	128
		6.3.8	Artificial Intelligence for Smart Education	129
	6.4	Existing Daily Lives Examples of Artificial Intelligence in Smart Cities		129
		6.4.1	Cameras	130
		6.4.2	Traffic and Parking System	130
		6.4.3	Drones	130
		6.4.4	Face Recognition	131
		6.4.5	Waste Management	131
		6.4.6	Autonomous Vehicles	131
		6.4.7	Spam Detection Software	132
		6.4.8	Recognition and Detection of Facial Expressions	132
		6.4.9	Navigation	133
		6.4.10	Games	133
		6.4.11	Automated Text Corrections and Text Editors	133
	6.5	Artificial Intelligence in Social Media for People in Smart Cities		134
		6.5.1	Instagram	134
		6.5.2	Facebook	134
		6.5.3	Twitter	134
		6.5.4	Marketing	135
		6.5.5	Chatbots	135
		6.5.6	Banking Apps	135
	6.6	Open Research Challenges and Research Directions		136
	6.7	Summary		137
	References			138
7	**Energy Management of Sustainable Smart Cities Using Internet-of-Energy**			143
	7.1	Introduction		143
	7.2	Recent Advances in Energy Sector		148
		7.2.1	Internet of Energy	148
		7.2.2	Energy 4.0	150
		7.2.3	Smart Grid	151
	7.3	Classification of Energy Management in Sustainable Smart Cities		153
		7.3.1	Scheduling Optimization	153
		7.3.2	Low Power Device Transceivers	154
		7.3.3	Cognitive Framework	154
		7.3.4	Cloud Computing Technology	154
	7.4	Internet of Energy Harvesting Things in IoE-Based Smart Cities		154
		7.4.1	Receiver Design	155
		7.4.2	Energy Optimization	155

		7.4.3	Energy Sources	155
		7.4.4	Energy Scheduling	156
		7.4.5	Energy Routing	156
	7.5	Green Energy		156
	7.6	Conceptual Framework of Proposed Energy-Based Quantum IoT Network Architecture for Sustainable Smart City		157
		7.6.1	Physical Layer	157
		7.6.2	Quantum Network Layer	159
		7.6.3	Quantum Teleportation Layer	160
		7.6.4	Application Layer	160
	7.7	Enabling Technologies in Energy Sector		160
		7.7.1	Internet of Things	161
		7.7.2	Artificial Intelligence	162
		7.7.3	Cloud/Fog/Edge Computing	162
		7.7.4	Quantum Computing	164
		7.7.5	Big Data	164
		7.7.6	Blockchain	165
	7.8	Open Research Challenges and Research Directions		165
	7.9	Summary		168
	References			168
8	**Internet of Vehicles for Sustainable Smart Cities**			175
	8.1	Introduction		175
	8.2	Internet of Vehicles Applications		177
		8.2.1	Transportation-Related IoV Applications	179
		8.2.2	Smart City-Related IoV Applications	179
	8.3	Connected and Autonomous (Level 0 to Level 5)		180
	8.4	Layered Architecture of Internet of Vehicles		183
	8.5	Internet of Vehicles Communication Model		184
		8.5.1	Vehicle-to-Vehicle (V2V)	184
		8.5.2	Vehicle to Roadside (V2R)	185
		8.5.3	Vehicle to Infrastructure (V2I)	186
		8.5.4	Vehicle to Home (V2H)	187
		8.5.5	Vehicle to Everything (V2X)	187
		8.5.6	Vehicle to Grid (V2G)	188
		8.5.7	Vehicle to Pedestrian (V2P)	188
	8.6	Open Research Challenges and Research Directions		189
	8.7	Summary		190
	References			190
9	**Smart Healthcare in Sustainable Smart Cities**			195
	9.1	Introduction		195
	9.2	Healthcare 4.0		199
	9.3	The Internet of Medical Things		200
	9.4	Smart Healthcare Services		201

	9.5	Enabling Technologies in Healthcare for Sustainable Smart Cities	205
		9.5.1 Internet of Things	205
		9.5.2 Healthcare Cloud, Fog, and Edge Computing	206
		9.5.3 Big Data	207
		9.5.4 Blockchain Technology	208
	9.6	Smart Healthcare Applications	209
	9.7	Open Research Challenges and Research Directions	213
	9.8	Summary	215
	References		216
10	**Unmanned Aerial Vehicles in Sustainable Smart Cities**		**221**
	10.1	Introduction	221
	10.2	Role of UAVs in Sustainable Smart Cities	223
	10.3	Technologies Used for UAVs in Sustainable Smart Cities	225
		10.3.1 Internet of Things (IoT)/Internet of Drone (IoD)	227
		10.3.2 Artificial Intelligence (AI)/Machine Learning (ML)	228
		10.3.3 Blockchain	228
		10.3.4 Wireless Sensor Networks	229
		10.3.5 Big Data	229
	10.4	Applications of UAV-Based Sustainable Smart Cities	230
		10.4.1 Smart Healthcare Services	231
		10.4.2 Security and Privacy	231
		10.4.3 Traffic and Crowd Monitoring	231
		10.4.4 Order Shipment	232
		10.4.5 Infrastructure and Environment Survey	232
		10.4.6 Disaster Management	233
	10.5	Open Research Challenges and Research Directions	233
	10.6	Summary	234
	References		234
Index			**239**

List of Figures

Fig. 1.1	Overview of sustainable smart city	4
Fig. 1.2	Preliminary data analysis: (**a**) documents published per year with the number of publications, (**b**) document per year by source with subject area, and (**c**) the congregation of co-appearing keywords from the Scopus database	5
Fig. 1.3	Concept of sustainable smart city	6
Fig. 1.4	Characteristics of sustainable smart cities	8
Fig. 1.5	Driving and guiding principles to create sustainable smart cities	12
Fig. 1.6	Contribution of IoT in building smart city	15
Fig. 1.7	Wireless technologies employed in different domains of a smart city	20
Fig. 1.8	Sustainable smart city concept	21
Fig. 2.1	Realization of Society 5.0 with its evolutions and involvement of emerging technologies	41
Fig. 2.2	Industrial Revolution to Society 5.0	46
Fig. 2.3	Anxious feeling of inhabitants regarding the realization: a reported survey	48
Fig. 2.4	The connection of Society 5.0 with UN SDGs with its components, enabling technologies, and different applications	51
Fig. 2.5	Potential research challenges with possible solutions for Society 5.0	52
Fig. 3.1	Enabling technologies employed in different domains of sustainable smart city	60
Fig. 3.2	Key enabling technologies for sustainable smart city	61
Fig. 3.3	Preliminary data analysis. (**a**) Documents published per year with the number of publications. (**b**) Document per year by source with subject area. (**c**) The congregation of co-appearing keywords from the Scopus database	63

Fig. 4.1	Concept of IoT in sustainable smart cities.	76
Fig. 4.2	Preliminary data analysis. (**a**) Documents published per year with the number of publications, (**b**) document per year by source with subject area, and (**c**) the congregation of co-appearing keywords from the Scopus database	77
Fig. 4.3	ICT architecture with IoT for sustainable smart city	79
Fig. 4.4	The security module of IoT architecture for sustainable smart cities	81
Fig. 5.1	Connection in 6G era integrating digital, physical, and biological world	98
Fig. 5.2	Evolution of generations from 1G to 6G	99
Fig. 5.3	6G-IoT framework for sustainable smart city	106
Fig. 5.4	6G-IoT communication system enabling technologies	108
Fig. 5.5	6G-IoT communication system use cases	111
Fig. 5.6	Challenges, open issues, and research direction for 6G.	114
Fig. 6.1	Recent utilization of artificial intelligence in sustainable smart cities	120
Fig. 6.2	Preliminary data analysis (**a**) documents published per year with the number of publications, (**b**) document per year by source with subject area, (**c**) the congregation of co-appearing keywords from the Scopus database	122
Fig. 6.3	Relation between artificial intelligence, machine learning, and deep learning	123
Fig. 6.4	Applications of artificial intelligence in sustainable smart cities	125
Fig. 7.1	Applications of smart cities in energy sector	144
Fig. 7.2	Concept of Internet of Energy	149
Fig. 7.3	Use cases for IoE in sustainable smart cities	150
Fig. 7.4	Applications of smart cities in smart grid	152
Fig. 7.5	Classification of energy management for IoT in smart cities	153
Fig. 7.6	Internet of Energy harvesting Things in IoE-based smart cities	155
Fig. 7.7	Proposed energy-based quantum IoT architecture for sustainable smart city	158
Fig. 7.8	Enabling technologies in energy sector	161
Fig. 7.9	Challenges in energy management-based sustainable smart city	166
Fig. 8.1	The Internet of Vehicles (IoV) concept	177
Fig. 8.2	Preliminary data analysis (**a**) documents published per year with the number of publications, (**b**) document per year by source with subject area, (**c**) the congregation of co-appearing keywords from the Scopus database	178
Fig. 8.3	Levels of connected and autonomous (level 0 to level 5)	181

Fig. 8.4	Internet of vehicle communication model	184
Fig. 9.1	Smart and connected healthcare system	197
Fig. 9.2	Preliminary data analysis (**a**) the published documents per year with the number of publications, (**b**) source with subject area, and (**c**) the congregation of co-appearing keywords from the Scopus database	198
Fig. 9.3	Healthcare 1.0 to 4.0	200
Fig. 9.4	Concept of Internet of Medical Things (IoMT)	201
Fig. 9.5	IoT paradigm in healthcare	206
Fig. 10.1	Applications of UAV in the sustainable smart cities	222
Fig. 10.2	Preliminary data analysis (**a**) documents published per year with the number of publications, (**b**) document per year by source with subject area, (**c**) the congregation of co-appearing keywords from the Scopus database	226
Fig. 10.3	Technologies used for UAVs in sustainable smart cities	227
Fig. 10.4	Application of UAV-based sustainable smart cities	230

List of Tables

Table 1.1	Definition of smart and sustainable smart cities	9
Table 2.1	Definitions of the Fifth Industrial Revolution	44
Table 2.2	Description of 17 United Nations Sustainable Development Goals	50
Table 5.1	Requirements and KPIs of 6G	105
Table 6.1	Reported literatures on artificial intelligence for smart cities components	125
Table 7.1	List of projects involved in cities	146
Table 10.1	Applications of UAV in various components of smart cities	225

About the Authors

Priyanka Mishra is currently pursuing a post-doctoral research fellowship with the Centre for Smart Information and Communication Systems in the Department of Electrical and Electronic Engineering Sciences, Auckland Park Kingsway Campus, University of Johannesburg, South Africa. She has authored more than 50 publications in various journals, conferences, and book chapters. She is a reviewer for several updated journals and conferences. She has supervised various master's thesis. Her area of interest includes next-generation communication systems (5G/6G), applications of 5G/6G in sustainable smart cities, industry 4.0/5.0, healthcare 4.0, intelligent transport systems (ITS), energy management (IoE), signal processing, and NOMA.

Ghanshyam Singh (Member, IEEE) received the Ph.D. degree in electronics engineering from the Indian Institute of Technology (Banaras Hindu University) Varanasi, Varanasi, India, in 2000. Currently, he is working as a Full Professor and Director at the Centre for Smart Information and Communication Systems, Department of Electrical and Electronic Engineering Science, Auckland Park Kingsway Campus, University of Johannesburg, South Africa. He was associated with the Central Electronics Engineering Research Institute, Pilani, and Institute for Plasma Research, Gandhinagar, India, respectively, where he was a Research Scientist. He also worked as an Assistant Professor with the Electronics and

Communication Engineering Department, Nirma University of Science and Technology, Ahmedabad, India. He was a Visiting Researcher with Seoul National University, Seoul, South Korea. He also worked as a Professor with the Department of Electronics and Communication Engineering, Jaypee University of Information Technology, Wakanaghat, Solan, India. He has more than 23 years teaching and research experience. His research and teaching interests include millimeter/THz wave technologies and their applications in communication and imaging, next-generation communication systems (5G/6G)/cognitive radio/NOMA, resource allocation and management, interference management, applications of 5G/6G in sustainable smart city, industry 4.0/5.0, healthcare 4.0, intelligent transport systems, energy management (IoE), and digital farming. He is the author/coauthor of more than 290 scientific papers of the refereed journals and international conferences and several books and book chapters published by Springer, Wiley, IET, CRC, and Academic press. He has supervised various MTech and PhD theses. He has worked as an editor/an associate editor and a reviewer for several reputed journals and conferences.

Chapter 1
Introduction: Importance of Sustainable Smart City

1.1 Introduction: Smart City and Its Demands

Recent rapid global urbanization trends and critical issues around sustainability present great challenges for the regular cities. The smart city concept has been established as a strategy for working with regular cities as they become systematically more complex through interconnected frameworks and increasingly rely on the use of information and communication technology (ICT) and Internet of Things (IoT) to fulfill the demand of their inhabitant. Thus, the smart cities are defined as the cities where investments in human and social capital, as well as traditional and modern communication infrastructure, drive sustainable economic growth and ensure a high quality of life. These cities exhibit intelligent management of natural and man-made resources through participatory governance. The concept of smart city is to create a liveable, sustainable, and prosperous society by utilizing state-of-the-art technologies across its various components such as waste management, mobility, water, energy, transport, healthcare, educations, etc. However, it also puts strain on city infrastructure [1]. Further, the strategic sustainable development (SSD) approach is applied as a method to exploit the benefits of the concept and to mitigate any identified limitations. Therefore, an emerging technology-based smart city is one of the best options to ensure efficient and sustainable resource management with security and satisfaction of inhabitant. Technologies such as IoT and ICT are integrated with a city's system to monitor, measure, and analyze the big data, generated by its users and stored in a protected environment, which could be used for a multitude of improvements to the infrastructure. Smart cities are urban areas that use technology to improve the lives of its citizens.

Smart City Origination

In the early 1990s, the concept of smart cities was introduced as an integration of regular cities with ICT infrastructure and smart components to stimulate the growth of cities. In general, the term "smart" refers to an array of technological and digital concepts and interventions, especially those related to ICT. Technology related to the 4IR (Fourth Industrial Revolution) seems to be gaining a lot of attention. As well as this technology-intensive interpretation, smart could also mean "intelligent" or "knowledge-intensive." "Technology" could also be interpreted as encompassing innovative approaches, techniques, processes, non-conventional interventions and scientific innovation. A smart city conversation often uses the word "city" in multiple ways. In this context, it could mean:

- Small towns and cities, as well as those in rural areas
- Metropolitan
- Upgrading transportation and connectivity
- New residential, private, commercial, precinct, or greenfield developments
- More effective, data-driven decision-making

Objectives and Opportunities
Sustainable smart cities include the following objectives and opportunities:

- Smart city initiatives need to take into account both the benefits and challenges when planning and implementing them.
- It aims to provide an interpretive framework regarding various aspects of smart cities and to assist in developing a common understanding of the concept.
- In order to ensure this approach is appropriate to the local context, it is necessary to highlight the realities of cities when planning and implementing such initiatives.

Smart City Benefits
Planning and implementing smart city initiatives can be improved by sharing experiences. Some of these could be:

- Better quality of life, cost saving, and secure communities
- Enhanced engagement between citizen and municipalities
- Reducing expenditure
- Improving efficiency and productivity
- Reducing efforts and working smarter
- Planning and implementation
- Helping with decision-making
- Knowledge-sharing platforms
- Monitoring and assessment
- Economic growth

Over the years, smart cities have developed new approaches to making sustainable human settlements through new urban approaches [2]. Therefore, intelligence makes the city more capable of responding to future urban challenges. Due to this, smart cities have gained attention all over the world, especially in densely populated areas. Today, IoT has created the realization of a smart city and community that is integrating ICT for better management of city resources. Citizens benefit from quality services and a higher standard of living. The health, water, surveillance, energy, parking, and the environment are among the services provided. The main goal of a

1.1 Introduction: Smart City and Its Demands

smart city is to utilize natural resources and services efficiently. ICT contributes to a number of aspects of society, including health quality, vehicle management, resource management, and smart city infrastructure. Recent developments in IoT technology have encouraged researchers to develop its potential applications [3] with respect to smart cities. It aims to spread awareness of the concept of a smart city, in exchange for data/information collection within IoT services in industries [4]. Many different studies have been conducted on various topics, such as monitoring the environment in smart cities [5] and living standards in a smart city that focuses on four urban conditions, including climate, transport, environment, and human flow [6]. Several researchers have proposed various theories about smart cities from a different perspective. According to [7], urban intelligence requires connections among social, physical, business, and information technology infrastructures. Based on these theories, a smart city is an urban environment managed by ICT that improves the city's daily functioning and the performance of individuals [8].

There are a variety of topics covered in existing literature to create a sustainable smart city, such as smart governance, smart transportation, smart infrastructure, smart healthcare, smart education, and smart security, as depicted in Fig. 1.1. The majority of these components are, however, discussed in numerous articles. In this chapter, we have prioritized the smart city concept by identifying available topics in this research. Several perspectives are taken into account in this chapter when examining the concept of smart cities from a technical, societal, and economic perspective. As part of our smart city design, we have also highlighted modern technologies. A number of published papers in this area explore the smart city concept from the perspective of technologies like artificial intelligence, blockchain, IoT, and computer technology. It develops a comprehensive overview of the fundamental challenges faced by smart cities in the development process.

- **IEEE Wireless Technology Standards in Smart City**

    The link between various devices is established via wireless technology. These IEEE standards—IEEE 802.11, IEEE 802.15.1, IEEE 802.15.3, IEEE 802.15.4, IEEE 802.15.6, and IEEE 802.16—are frequently utilized for wireless technology in the smart cities. IEEE 802.11 can be utilized in a variety of smart city applications, including smart waste management, smart homes, and smart transportation. In contrast, the coverage of IEEE 802.15.1, IEEE 802.15.3, IEEE 802.15.4, and IEEE 802.15.6 is noticeably less than that of IEEE 802.11. These standards are also more suited for smart city applications like smart lighting and smart health monitoring. The IEEE 802.16 standard specifies a number of long-range communication-supporting technologies. The smart grid, one of the potential applications of IEEE 802.16, can utilize the defined technologies of smart cities.

    In preliminary data analysis, Fig. 1.2a illustrates the number of documents published per year with the number of publications. The most publications are made in the year 2021 with more than 100, and as publications are growing each year, this is an excellent field for further research. Articles are the document type with a large number of publications with a percentage of 54.6%. Figure 1.2b shows the

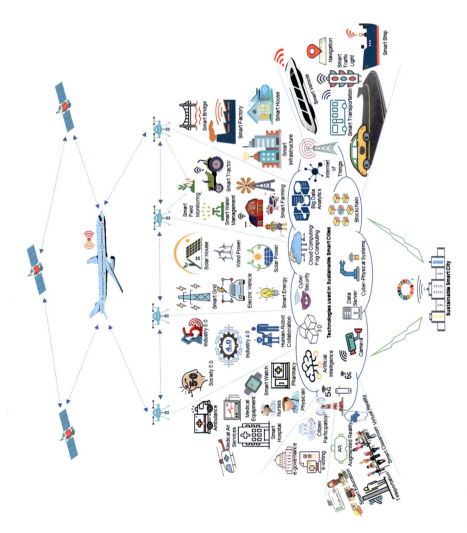

Fig. 1.1 Overview of sustainable smart city

1.1 Introduction: Smart City and Its Demands 5

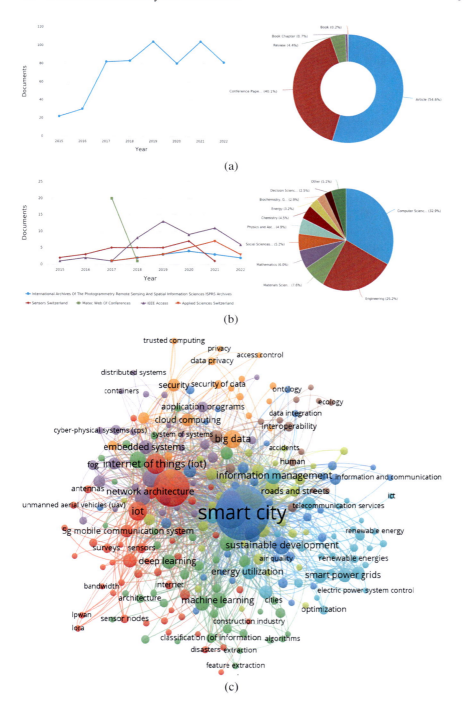

Fig. 1.2 Preliminary data analysis: (**a**) documents published per year with the number of publications, (**b**) document per year by source with subject area, and (**c**) the congregation of co-appearing keywords from the Scopus database

document per year by source with subject area. The MATEC Web of Conferences has publications of more than 20. Computer Science subject domain has most of the publications with a percentage of 32.9%, and Engineering has the second most publications with a percentage of 25.2%. Figure 1.2c shows the congregation of keywords that are occurring together in a paper. The interrelations between keywords that are co-appearing in the research are linked together. Smart city, sustainable development, Internet of Things, big data, and machine learning are the most co-appearing keywords.

The rest of the chapter is organized as follows. Section 1.2 explains the sustainability of smart cities. The definition of a smart city and a sustainable smart city is presented in Sect. 1.3. Section 1.4 explains driving and guiding principles to create sustainable smart cities. Various potential components of the sustainable smart city are analyzed and defined in Sect. 1.5, including smart governance, smart transportation, smart energy, smart healthcare, smart industry, smart technology, smart building and infrastructure, smart agriculture, and smart education. The requirements and key performance indicators are presented in Sects. 1.6 and 1.7, respectively. Smart cities around the world are presented in Sect. 1.8. Open research challenges and research directions are discussed in Sect. 1.9. Organization of the book is given in Sect. 1.10. Finally, Sect. 1.11 concludes the chapter with a summary.

1.2 Sustainability in Smart City

Smart and sustainable cities have fully integrated utility networks that can adjust in real time to meet the expected demand as depicted in Fig. 1.3. The smart and sustainable city is an interconnected system, in which each system is autonomous and interdependent. The interconnected nature of the citys' transportation, energy, healthcare, manufacturing, education, hospitality, agriculture, and utilities system would function if the city were subjected to a natural disaster. When the entire city receives alerts about the approaching disaster, this information is shared across various sectors. The transportation system adjusts routes and schedules to ensure the safe evacuation of residents and the movement of emergency personnel. The energy system implements measures to mitigate power outages and ensure the availability of backup generators. Healthcare facilities prepare for an influx of patients, while manufacturing facilities switch to producing essential supplies and equipment. Educational institutions adapt their operations to provide remote learning and support services. The hospitality industry collaborates with emergency response teams to provide temporary shelters and aid to those affected. Agricultural systems secure crops and livestock, while utilities prioritize the safety and functionality of critical infrastructure. In addition, train

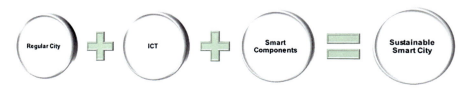

Fig. 1.3 Concept of sustainable smart city

1.2 Sustainability in Smart City

schedules are adjusted, and rolling stock is deployed as needed. The schedules for river and maritime transportation are also altered, as well as the locations of arrival and departure. There is a great deal of potential for integration, interdependency, and interoperability in healthcare, which would greatly improve the level of service citizens receive. An intelligent and sustainable city's citizens receive a continuous flow of information and advice about their healthcare needs, where their scheduled services have been moved, and how to plan their transportation. Through this interconnected approach, the city can effectively respond to the disaster, minimize its impact, and support the well-being of its residents.

Electricity and water, for example, are rerouted algorithmically if one network fails, so as not to affect the other. Different modes of transportation—air, surface, rail, and maritime—are interoperable and interdependent in their performance on the transportation network. By using intelligent control algorithms, traffic monitoring and transportation units maximize transportation flow across cities. The energy network integrates all energy sources, including carbon and alternative sources, to maximize their efficiency and minimize disruptions. In all energy-dependent systems, smart sensors allow for optimal energy consumption based on environmental information. The communication networks operate continuously, acquiring, interpreting, disseminating, and storing data and information. Data is cloud based, and cloud access is ubiquitous, from streets to parks to transportation. A sustainable smart city, envisioned above, does not require independent resilience assessments. In order to achieve seamless recovery of critical infrastructure, autonomous and interdependent systems and their optimized interactions produce resilience. For this, we need to focus on few characteristics of a city. The six main characteristics of development sustainable smart city are the following as shown in Fig. 1.4.

- **Smart Economy**

High-quality universities and advanced research institutes, as well as an advanced telecommunications infrastructure, contribute to the creation of employment and the advancement of advanced technology.

- **Smart Environment/Natural Resources**

Resources are intelligently used to promote sustainable development based on recycling and waste reduction, rational building criteria, and green space protection and management.

- **Smart Governance**

Cities can involve their citizens in public issues by adopting policies to promote territorial development and inter-municipal networking. In this way, it will be able to promote awareness and use technology to simplify and digitize administrative processes.

- **Smart Living (Quality of Life)**

A city can enhance its own tourist image with intelligent online promotions (city routes and thematic maps) by offering advanced services for improving quality of life (home care, childcare, aged care).

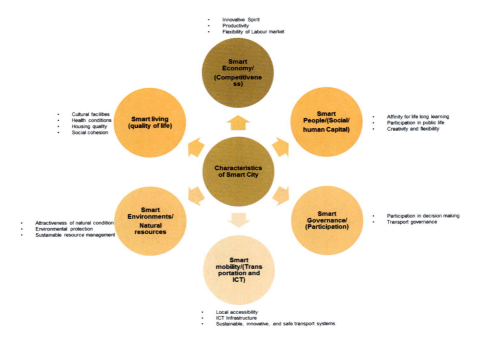

Fig. 1.4 Characteristics of sustainable smart cities

- **Smart Mobility**

A modern and efficient public transportation system in a city would allow people to easily travel from one place to another, thereby encouraging the use of environmentally friendly vehicles. Additionally, it would help regulate access to historic town centers and make them more livable by promoting pedestrian walkways.

- **Smart People**

When citizens of a city are active and participate in public life, a city can maximize its social capital and foster peaceful cohabitation.

1.3 Definition of Smart and Sustainable Smart City

A smart sustainable city is a concept that connects smart cities and sustainable cities. This is a relatively new phenomenon that became widespread around the middle of the decade (2010s) [9–11]. The concept arose out of five distinct developments, including smart cities, sustainable urban development, sustainable cities, sustainability and environmental challenges, and urbanization/urban growth [11]. Various publications all define and explore the different concepts of the smart city, sustainable city, and smart sustainable city. There are different definitions of smart cities [12, 13], and Table 1.1 lists the definitions of smart cities and sustainable smart cities that have been studied in the literature. The authors of [14] stated, "we believe

1.3 Definition of Smart and Sustainable Smart City

Table 1.1 Definition of smart and sustainable smart cities

Sr. no.	References	Definition (Smart City)
1.	[14]	"A city is smart when investments in human and social capital and traditional (transport) and modern information and communication technology (ICT) infrastructure fuel sustainable economic growth and a high quality of life, with a wise management of natural resources, through participatory governance"
2.	[20]	"Key conceptual components of a Smart City are three core factors: technology (infrastructures of hardware and software), people (creativity, diversity, and education), and institution (governance and policy). Given the connection between the factors, a city is smart when investments in human/social capital and information technology (IT) infrastructure fuel sustainable growth and enhance a quality of life through participatory governance"
3.	[21]	"A city well-performing in a forward-looking way in economy, people, governance, mobility, environment, and living, built on the smart combination of endowments and activities of self-decisive, independent, and aware citizens"
4.	[22]	"A Smart City is based on intelligent exchanges of information that flow between its many different subsystems. This flow of information is analyzed and translated into citizen and commercial services. The city will act on this information flow to make its wider ecosystem more resource-efficient and sustainable. The information exchange is based on a smart governance operating framework that is designed to make cities sustainable"
5.	[15]	"A Smart City [refers to] a local entity—a district, city, region or small country—that takes a holistic approach to employ information technologies with real-time analysis that encourages sustainable economic development"
6.	[23]	"The application of information and communications technology (ICT) with their effects on human capital/education, social and relational capital, and environmental issues is often indicated by the notion of a Smart City"
7.	[24]	"Smart city is a city that requires a proper governance system for connecting all forces at work, allowing knowledge transfers, facilitating decision-making in order to maximize their socio-economic and environmental performance"
8.	[25]	Smart city is a city that uses information communication technologies to enhance the quality of life for residents, the use of modern technologies to increase efficiencies in order to better utilize available resources, the strive for increasing sustainability in order to meet commitments set by international organizations and the fostering of a culture of creativity and innovation that attracts educated workers to further create knowledge-based solutions to a city's challenges"
9.	[26]	Smart city is a city that achieves its targets, such as security, mobility, scalability, latency and deployment"

(continued)

Table 1.1 (continued)

Sr. no.	References	Definition (Sustainable Smart City)
1.	[27]	"A Smart Sustainable City is one in which the seams and structures of the various urban systems are made clear, simple, responsive, and even malleable via contemporary technology and design. Citizens are not only engaged and informed in the relationship between their activities, their neighborhoods, and the wider urban ecosystems, but are actively encouraged to see the city itself as something they can collectively tune, such that it is efficient, interactive, engaging, adaptive, and flexible, as opposed to the inflexible, monofunctional, and monolithic structures of many twentieth century cities"
2.	[28]	"A smart and sustainable city invests in human and social capital, manages resources wisely, has citizens who participate in city governance, and has a traditional and modern infrastructure that supports economic growth to create a high quality of life for its inhabitants"
3.	[29]	"Cities that use ICT technologies to be more intelligent and efficient in the use of resources, resulting in cost and energy savings; improved service delivery and quality of life; and a reduced environmental footprint"
4.	[30]	"The Smart Sustainable City seeks to achieve concern for the global environment and lifestyle safety and convenience through the coordination of infrastructure. Smart Sustainable Cities are realized through the coordination of infrastructures consist of two infrastructure layers that support consumers' lifestyles together with the urban management infrastructure that links these together using IT"
5.	[31]	"A city that strategically utilizes many smart factors such as information and communication technology to increase the city's sustainable growth and strengthen city functions, while guaranteeing citizens' happiness and wellness"
6.	[32]	"To expose the feasibility of rapidly progressing toward our energy and climate objectives at a local level while proving to citizens that their quality of life and local economies can be improved through investments in energy efficiency and the reduction of carbon emissions"
7.	[33]	"Must provide ICT-based solutions to address public issues as well as to improve competitiveness to ensure a more sustainable future of the city"
8.	[34]	"A SSC must deliver a sustainable, prosperous, and inclusive future for its citizens through an effective integration of its digital, physical, and human systems"
9.	[35]	"A smart sustainable city is an innovative city that uses ICTs and other means to improve the quality of life, efficiency of urban operation and services, and competitiveness, while ensuring that it meets the needs of present and future generations with respect to economic, social, and environmental aspects"
10.	[11]	"A city that meets the needs of its present inhabitants, without compromising the ability for other people or future generations to meet their needs, and thus does not exceed local or planetary environmental limitations, and where this is supported by ICT"

a city to be smart when investments in human and social capital and traditional (transport) and modern (ICT) communication infrastructure fuel sustainable economic growth and a high quality of life, with a wise management of natural resources, through participatory governance." Numerous definitions of "smart cities" are still lacking in terms of sustainability. The application of this idea to urban planning has resulted in the creation of a number of sustainable city models that reflect various aspects of sustainability, including "zero-carbon eco-cities" and "ubiquitous eco-cities" [15]. These city plans, especially from the economic and social perspectives, are far away from the true concept of a sustainable city if they are evaluated according to whether they achieve the objectives of all three sustainability pillars [16]. Culture is becoming a fourth pillar of sustainable development in addition to the three previously mentioned pillars. Urban planning is said to benefit from culture because it fosters "thoughtfulness and openness and contributes to a world with complementary, pluralized development visions" [17]. According to the sustainability approach, this fourth pillar should be included in the creation of smart cities; for this reason, it's crucial to combine the sustainability and smart city frameworks to take both points of view into account. As a result, "smart sustainable cities" should be used instead [18].

The term "sustainable smart city" is used by authors in [19] to describe a city that is "supported by a pervasive presence and large usage of modern ICT, which, in conjunction with diverse urban domains and systems and how these intimately interrelate, enables cities to become more sustainable and to provide residents with a better quality of life" [19]. Based on all of the constituent definitions, we find that there is consensus on the importance of ICT in smart urban development and that the definition of a smart (sustainable) city is broad and has a holistic point of view where people and the environment are given consideration.

1.4 Driving and Guiding Principles to Create Sustainable Smart Cities

A smart city must follow by six interrelated principles in order to be inclusive. When decisions need to be made about the identification, planning, and implementation of smart initiative and technologies, these principles offer guidance. Each of the guiding principles is expressed as a goal that should be attained to improve the inclusiveness of a smart city effort. A smart initiative or technology's nature and purpose should be determined by the following guidelines:

- Smart for everyone and everywhere
- Technology as an enabler
- The local context
- Needs of the city and community
- Innovation, alliances, and participation
- Sustainable, secure, and robust

The following sections provide practical guidance on how to implement the principles for planning and implementing smart cities is depicted in Fig. 1.5.

Fig. 1.5 Driving and guiding principles to create sustainable smart cities

1.4.1 Smart for Everyone and Everywhere

Smart cities should integrate cutting-edge technologies and initiatives that collectively improve a city's well-being. A smart city initiative should not be undertaken at the expense of certain parts of the city or society. Consequently, a smart city should ultimately benefit all its residents. The use of smart initiatives is not limited to individuals with financial resources or a comprehensive understanding of technology. Smart city initiatives should be planned and implemented in an innovative way to ensure they meet the criteria of smart cities. Ensuring that all sectors of society can access and benefit from smart initiatives is imperative during the planning and implementation process, even for those with limited financial mean.

1.4.2 Technology as an Enabler

In enabling and improving various functions and operations of the sustainable smart cities, information and communication technologies play a very significant role. By employing an emerging technology, the service providers could provide services to the cities and society more effectively and efficiently. Instead of assuming that technology will automatically solve a particular problem, real needs and requirements must be considered when determining which technologies to choose and implement. Technology should be selected based on the local context, needs, and resources. It was developed to refer to the era of cyber-physical systems that extend beyond mere automation. This era encompasses decentralized industries and systems that are transparent, integrated, self-optimizing, self-configuring, and self-diagnosing. These advancements are commonly referred to as the Fourth Industrial Revolutions (4IR) or Industry 4.0. Technologies and concepts that are commonly associated with 4IR and smart cities include the IoT, human enhancement technologies, virtual reality and augmented reality, near-field communication, advanced materials and smart materials, speedy connectivity (5G and Wi-Fi), 3D printing and additive manufacturing, big data, distributed ledger technologies and blockchains, smart electrical grids, bots, drones, satellite enablement, facial recognition, and autonomous or driverless vehicles.

1.4.3 The Local Context

Smart cities require careful consideration of local context when planning technologies and initiatives. There are a number of factors common to cities and communities, such as poverty, inequality, and unemployment.

1.4.4 Needs of the City and Community

Active community participation could contribute significantly to the success of smart city initiatives, including identifying, developing, and implementing them. In a participatory process, people take an active role in decision-making from the outset. In addition to participating in the planning and design phases, they should also help with the execution and management of the city. In general, the term city refers to all relevant stakeholders and affected parties. Residents, businesses, universities, research organizations, government departments, and industry are all part of this community. Another facet of community engagement is the contribution that the community can make once an initiative has been adopted. They could provide feedback to improve the initiative, they could contribute by sharing information and data, and they could help with monitoring and evaluation processes.

1.4.5 Innovation, Alliances, and Participation

It is important for a smart city to include both public and private initiatives, innovation, alliances, and participation. It is essential to find reliable partners that share the same values and objectives. There may be different motivations for partners to participate in a smart city initiative. It can sometimes be challenging to reconcile, for instance, purely commercial intentions of some partners with benevolent intentions of others. Managing alliances is crucial, and roles and expectations must be clearly understood by all partners. It is true that sometimes the private sector drives smart city agendas, rather than the government. In spite of this, local governments must ensure smart city initiatives are guided by appropriate urban development and planning policies, strategies, plans, and frameworks.

1.4.6 Sustainable, Secure, and Robust

It is essential for cities to be inclusive, sustainable, safe, and robust. Therefore, smart city initiatives must consciously incorporate principles of sustainability, resilience, and safety. The human, social, economic, and environmental dimensions of sustainable development can be addressed through smart interventions and technologies. By creating economic opportunities and addressing environmental challenges, smart cities can improve the quality of life of all residents. An integrated, multifaceted approach is needed to improve the sustainability of a city. The use of smart technologies can contribute to the creation of safe communities. By improving the effectiveness and efficiency of scarce resources, they could assist the police and other law enforcement agencies in preventing and detecting crimes. To improve city safety and security, private security and community groups could also utilize smart technologies and initiatives. It is possible to enhance the safety and security of communities using technologies and smart concepts such as CCTV surveillance, artificial intelligence, facial recognition, predictive policing, data, community safety apps, and e-policing.

1.5 Potential Components of Sustainable Smart City

The potential components of sustainable smart cities are smart government, smart transportation, smart energy, smart healthcare, smart industry, smart technology, smart building and infrastructure, smart agriculture, and smart education. These components assist in dealing with various aspects such as participation in decision-making, services, governance, flexibility, creativity, healthcare, education, ICT infrastructure, sustainable, innovation, transport systems, entrepreneurship, environmental protection and sustainable resource management [36, 37]. In this section, we discuss some of the components as shown in Fig. 1.6.

1.5.1 Smart Governance

Governments can leverage social media to encourage people to contribute and work in smart cities. In addition, smart governments should not only advance in the pursuit of technological development but also have sound governance and wise policies [36]. According to [38], social media can be a viable marketing tool. However, the government must have a role in this endeavor. A smart city's success depends on providing reliable services to its citizens. In order to promote resource efficiency, cloud-based information system services are necessary. By collaborating with government organizations and stakeholders, information systems and related services

1.5 Potential Components of Sustainable Smart City

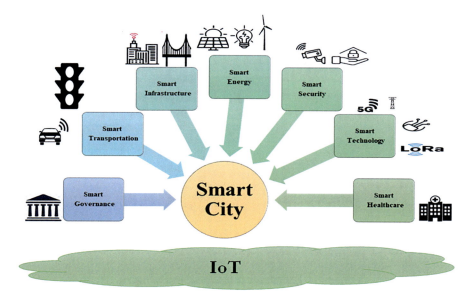

Fig. 1.6 Contribution of IoT in building smart city

made available to citizens. Building associations enhances the effectiveness of the intelligent governance approach in any area [39].

1.5.2 Smart Transportation

As long as humankind has existed, transportation has been a necessity. Road transportation, rail transport, water transport, and air transportation have increased in demand due to technological advancements. The modern transportation system is integrated with a variety of navigation and communication systems [40–42].

Additionally, smart travel systems provide passengers with information related to traffic congestion levels, alternative routes, etc. Moreover, smart travel systems enforce safety and security measures for pedestrians and passengers. Likewise, modern systems offer global air transportation, intelligent road networks, intersection trains, underground rail and municipal networks, community connected safety, and pedestrian-protected areas.

In [40], an algorithm was proposed for detecting air traffic conflicts in order to prevent accidents. This algorithm relies on the exchange of flight information for air traffic management. By providing real-time directions for travel speeds, [43] suggests the management of railway systems to increase time and fuel efficiency. Similarly, experts also worked on governing overcrowding and conflicts in railway networks.

Several studies have been conducted on road management [44] and road safety. Additionally, upcoming transport applications offer RFID-enabled toll collection, passport control at airports, parking management, and taxi hiring and follow-up via mobile applications. Since cars with global positioning systems (GPS) devices and cell phones are familiar with all drivers, many routes rely on GPS information to monitor drivers' location and traffic model [45]. This real-time information has been used for mapping in applications like Google Maps, etc. and is being used for public transportation planning. Drivers can also be directed to a nearby free parking location by feature parking systems.

1.5.3 Smart Energy

The typical electrical system uses one-way energy flow from a main generator source, usually a hydroelectric or fossil fuel-based plant. Power generation is controlled via feedback from the substations. Nevertheless, since the consumer end does not provide any feedback, the power generation scheme used in these systems requires that the power produced by these systems exceeds the demand by a substantial amount in order to ensure a continuous supply [46, 47]. In such systems, detecting faults and correcting them is also a time-consuming process. Moreover, as renewable energy technologies become more affordable, consumers today not only receive a supply from the main utilities but also produce their own power. By using ICT technologies, smart grids makes current and newly installed grids more observable, provide distributed energy generation at both consumer and utility ends, and introduce self-healing capabilities. The smart grid transmits real-time power data from different points on the grid to utilities throughout the supply line to customers. Since smart grids provide real-time data about consumer usage, they make it easier to manage power generation by developing prediction models based on consumption data acquired, integrating a variety of energy sources, as well as self-healing [48] to ensure an uninterrupted power supply.

1.5.4 Smart Healthcare

Due to high population growth, traditional healthcare is surviving. With a growing population and limited resources, conventional health systems need to be innovative, efficient, and sustainable. Smart biosensors, wearable gadgets, and ICTs are all part of intelligent healthcare. Many components of smart healthcare involve the emergence of body sensors, intelligent hospitals, and quick response in emergency. Modernized healthcare services use cloud computing, sensor networks, smartphone applications, ICT, and data processing methods to meet the needs and raise the standard of service [49]. An IoT-enabled health system was developed [50] to accurately diagnose and monitor patients, and biomedical equipment. The quality of healthcare services contributes to a higher standard of living for urban residents. A major improvement in the awareness of smart cities is the integration of smart healthcare into smart cities.

1.5.5 Smart Industry

Across the globe, industries are striving to become more efficient and productive while reducing costs. Among the key components of Industry 4.0 is the concept of a connected factory, where all intermediaries are seamlessly integrated and work together; IoT is responsible for making this possible [51]. The use of IoT in manufacturing and production processes, as well as cyber-physical systems integrating workers and machines, has contributed to a number of industry benefits, including faster and better innovation, optimization of manufacturing schemes (resources and processes), better-quality products, and enhanced worker safety. Using IoT in smart industries, however, comes with several challenges. Working with heterogeneous devices and machines presents its own challenges. IoT applications for smart industries require flexible configuration, connectivity, and fast implementation of cyber-physical systems [52]. The development and deployment of Industry 4.0 services has been spurred by artificial intelligence and the IoT. Data from embedded sensors in machines and other factory processes is used to automate, perform business intelligence operations, and increase automation. Researchers have developed frameworks for integrating artificial intelligence into IoT for smart industries [52–54]. In the manufacturing industry, artificial intelligence is primarily used for predictive maintenance, monitoring, and fault detection (machine health).

1.5.6 Smart Technology

In order to build, utilize, and operate smart cities, smart technology is essential [55]. In smart cities, various technologies are used, including 5G, ZigBee, Bluetooth, LTE A, VANET, RFID, LPWAN, LoRA, and Wi-Fi. In many cases, the technical aspect is ignored despite the fact that it might play a role in the solution to another issue as well. A connected, digital, and virtual city is created by combining technology with cutting-edge solutions [56]. The city needs to use existing technology to improve residents' satisfaction with its performance.

1.5.7 Smart Buildings and Infrastructure

In the smart infrastructure world, a smart power grid, also known as a smart grid, is a type of smart infrastructure. Smart grids enable the two-way flow of electricity and data between utility companies and consumers. This allows for more efficient management of electricity resources and helps reduce energy waste. Smart grids also enable users to track and manage their electricity usage in real time. As part of the building and infrastructure, there are various components including energy sources, performance control systems, load balancing methods, intelligent meters, fault-tolerant mechanisms, and operational control mechanisms. The objective of implementing green structures is to enhance energy efficiency and reduce carbon emissions. In the context of a smart home, monitoring household items, energy consumption, surveillance, lighting controls, and other aspects can contribute to improving citizens' quality of life. Additionally, waste management is recognized as a crucial element in building a smart city/society in the present world [57]. Because cities and industries are expanding, garbage production is exponentially growing. The waste management system controls the handling of man-made waste, the services of urban areas, and the needs of private offices [58]. There are few key categories of waste management, namely, waste collection, waste disposal, and waste recycling. It is imperative for smart cities to manage waste properly, because improper waste disposal puts both human and environmental health at risk.

1.5.8 Smart Agriculture

Among the Sustainable Development Goals for 2030, food security is of the utmost importance. In countries around the world, the race for sustainable food production and the efficient use of dwindling resources, such as water, has been a top priority. This is because the world population grows and climate change worsens, causing erratic weather in food centers. Smart agriculture involves embedding sensors into plants and fields to measure various parameters to assist in decision-making and prevent diseases, pests, etc. [59]. Precision agriculture is part of the smart agriculture paradigm, in which sensors are placed in plants to provide targeted measurements and thus allow targeted care mechanisms to be implemented. As part of the fight for sustainable food production, precision agriculture will be essential for food security in the future. A major application of artificial intelligence in IoT for agriculture is crop monitoring/disease detection, as well as data-driven crop care and decision-making [60].

1.5.9 Smart Education

In order for smart cities to function effectively, education is crucial. For both adults and children, education may be the most significant smart city service. The only way to remain competitive in a rapidly globalizing world is to develop and continue to build knowledge-based skills through education. In addition to initial knowledge (such as that acquired through schooling, vocational training, and university education), lifelong learning is also considered. Therefore, schools and universities must be considered when designing smart education solutions [61].

1.6 Requirements of Smart City

IoT-based "smart cities" require advancements in wireless technology and revolutionary innovations in technology to address the constraints, challenges, and applications of next-generation wireless networks. Among the requirements set by the International Telecommunication Union (ITU) for 5G are 20 Gbps per cell up to 1 Gbps per user, sub-millisecond latency, and the ability to serve more than one million devices per square kilometer. While 5G deployments claim to have achieved 1 Gbps download speeds, the average download and upload rates in typical real-world scenarios are 200 Mbps and 100 Mbps, respectively. Today, latency in real-world situations is between 21 and 26 milliseconds, but it is expected to fall below 1 millisecond in the future. The true value of 5G will be realized in automation, smart cities, IoT applications, and more. The system must have extremely high capacity (greater than 100 Gbps), low latency, massive devices, high coverage, high reliability, low energy, long battery life, and low cost. A fifth-generation mobile communication system replaces MIMO (multiple-input multiple-output) with massive MIMO and centralized systems with distributed systems, i.e., D2D communications. When there are more antennas, the capacity of the channel increases proportionally. Several hundred or even thousands of antennas are used in massive MIMO. In addition to improving capacity, diversity gain, and bandwidth, more antennas offer capacity gain and diversity gain. Over each transmit antenna, the same data stream is transmitted orthogonally or nearly orthogonally. It exploits multipath fading to its advantage, thereby improving the reliability of the received signal. The diversity gain increases as the number of transmitting and receiving antennas increases. Through beamforming, a large number of antennas can direct a signal to narrow regions in space, mitigating the effects of interference.

In general, smart cities are divided into two parts: the requirements for the smart city and the specifications needed to meet these requirements. As the data rate for the smart city is essential, the average rate must range between 10 and 20 Gbps, as specified by the ITU for the fifth generation, i.e., almost 100 Gbps. This is determined by the closeness to the Shannon capacity, i.e., maximum achievable capacity in an error-free environment. Therefore, how close we can get to Shannon to achieve a data rate that is compatible with 5G specified by ITU depends on the signal-to-noise ratio and the bandwidth. Secondly, there are distributed systems, from a base station to D2D or M2M communication, that is, two devices can operate independently without a base station.

By replacing heavy base stations, power will be transferred from the base station to devices. Power is being transferred from the centralized system to distributed systems. Now, in specifications, we have power optimization and reliability. The link must be reliable to provide a degree of freedom. If there is an error in one link, we should be able to use another link. With the goal of maximizing energy utilization, power optimization techniques play a crucial role in smart cities. Power optimization refers to power efficiency, which entails the fair sharing of power between users, ensuring equitable distribution even in scenarios of very low, finite, and

20 1 Introduction: Importance of Sustainable Smart City

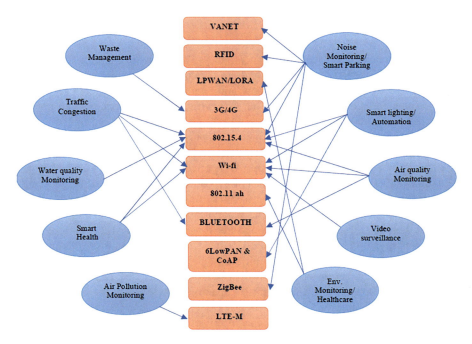

Fig. 1.7 Wireless technologies employed in different domains of a smart city

long-lasting battery power. Additionally, smart cities rely on a diverse range of wireless technologies, including VANET, RFID, LPWAN, LoRA, Wi-Fi, Bluetooth, ZigBee, LTE A, and 5G, as illustrated in Fig. 1.7.

1.7 Key Performance Indicators of Sustainable Smart Cities

In a smart sustainable city, ICTs and other means are used to improve the quality of life, the efficiency of urban operations and services, and the competitiveness of the city while simultaneously fulfilling the economic, social, environmental, and cultural needs of current and future generations [62–64] as depicted in Fig. 1.8.

1.7.1 Economic Sustainability

In order to evaluate the sustainability of the sustainable smart city economy, we should consider the following topics: ICT infrastructure, innovation, employment, trade, productivity, physical infrastructure, and public sector. A study should be conducted to discover whether sustainable smart city boosts local economies.

1.7 Key Performance Indicators of Sustainable Smart Cities

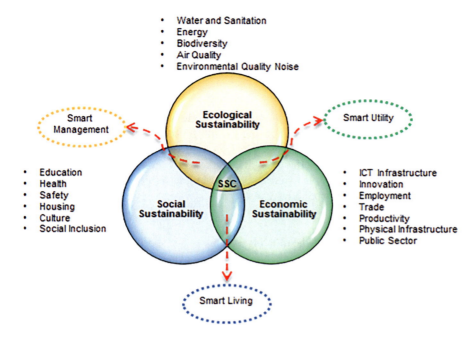

Fig. 1.8 Sustainable smart city concept

- **ICT Infrastructure**

  Intelligent and sustainable cities are enabled by ICT infrastructure, which is the foundation for other ICT solutions. In addition to terminals, networks, and access, ICT infrastructure also includes services and information platforms. It is imperative to take into account the use of various terminals (computer, mobile phone, tablet, etc.), the internet, wireless and fixed broadband, backbone networks, cloud computing platforms, and data centers when measuring ICT infrastructure.

- **Innovation**

  In order to evaluate a city's innovation capacity, multiple perspectives must be considered. The most innovative cities are those that can adapt to changes quickly and play a leading role in their region. Research and development investments are directly measured as well as the research and development output, which is well demonstrated by patents.

- **Employment**

  A city's employment rate is a key indicator of its economic health. It is important to note that in a city, both formal and informal employment exist, with formal employment referring to jobs within established organizations and informal employment encompassing self-employment, small-scale businesses, and other non-traditional work arrangements.

- **Trade**

  There is no doubt that trade is the foundation of commercial prosperity. It is possible to measure trade by looking at exports and imports. Additionally, e-commerce plays a significant role in sustainable smart cities' trade.

- **Productivity**

  In order to achieve sustainable smart cities, productivity needs to be assessed in terms of the use of information and media, the innovation of products and processes, and the leadership of businesses and services. This is because the success of a smart city depends on how effectively it uses data, how innovative its products and services are, and how it is able to attract businesses and services to the city. All these factors contribute to the productivity of the city.

- **Physical Infrastructure**

  Water supply, electricity, health infrastructure, transportation, road infrastructure, buildings, and urban planning, among other categories, should be improved in a city to make it smarter and more sustainable. All of these categories are essential for a city to function properly and to have a good quality of life for its citizens. Improved water supply and electricity will enable more efficient use of resources, while improved health infrastructure, transportation, road infrastructure, buildings, and urban planning will help create a more efficient and equitable urban environment.

- **Public Sector**

  The public sector is a part of the economy that provides governmental services. It typically includes government-run businesses, such as schools, hospitals, and public transportation, as well as social services like welfare and unemployment benefits. These services are provided to the public by the government, often with taxpayer money. Sustainable smart cities should prioritize the use of ICT for improving the efficiency of these services. By utilizing ICT, public sector services can become more user-friendly, cost-effective, and efficient. This results in improved quality of life for citizens and make the services more sustainable in the long run. Additionally, ICT helps reduce government bureaucracy and increase transparency, leading to better accountability in the public sector.

1.7.2 Ecological Sustainability

Environmental sustainability in sustainable smart cities is measured by considering the following six categories: air quality, water and sanitation, noise, environmental quality, biodiversity, and energy.

1.7 Key Performance Indicators of Sustainable Smart Cities

- **Air Quality**

  In many cities, air quality plays a major role in the quality of life. The main concern of city residents is air pollution, which should be accurately monitored and its data made publicly available. Another aspect of air quality is the CO_2-e emissions of the city, where "-e" represents equivalents and every other greenhouse gas is converted to CO_2.

- **Water and Sanitation**

  As part of environmental protection, water should be considered in the following aspects: water resources, water distribution, water conservation, wastewater treatment, drainage, and sanitation.

- **Noise**

  The level of noise exposure in the city is taken into account in this category. Noise levels are an important factor when determining the quality of a city's environment. High levels of noise exposure leads to a variety of health problems, including hearing loss, stress, and sleep disturbances.

- **Environmental Quality**

  In the following aspects, solid waste, electromagnetic fields, green areas, and public areas, the quality of the environment assessed through quantitatively or qualitatively. Solid waste is measured by the total amount generated and the rate of recycling. Electromagnetic fields are measured based on the level of radiation. Green areas are assessed by considering the size of the area and amount of vegetation present. Public areas can be measured in terms of accessibility and the quality of the facilities.

- **Biodiversity**

  There are many challenges involved in measuring biodiversity. One of the main challenges is that it is hard to define what exactly constitutes biodiversity. Additionally, it is difficult to measure the number of species in a given area, since species are very small or hard to detect. Additionally, it is hard to compare biodiversity between different areas since different areas may have different species. A number of aspects, such as native species and the natural environment for protecting them, can be included in the city plan.

- **Energy**

  Energy use in the city is included in this category, which includes electric consumption, renewable energy consumption, and household energy conservation measures. Electric consumption is the energy used to power all of the electrical appliances in a city, while renewable energy consumption is the energy sourced from natural resources such as solar and wind power. Household energy conservation measures are initiatives taken to reduce the amount of energy used in a household, such as installing energy-efficient appliances or using more efficient lighting systems.

1.7.3 Social Sustainability

As far as society and culture in a sustainable smart city are concerned, performance in these six domains should be considered: education, health, safety, housing, culture, and social inclusion.

- **Education**

  The importance of education and training cannot be overstated when it comes to enhancing human creativity and improving human resources. Education and training provide the knowledge and skills needed to be creative and productive. They help to develop critical thinking, communication, problem-solving, and other skills that are essential for success in the modern world. It is possible to assess education improvement from a variety of perspectives, such as education investment, ICT usage as an aid, student capability improvement, adult literacy, etc.

- **Health**

  It is generally accepted that healthcare and medical services fall under the category of health services. Healthcare and medical services are typically used to diagnose, prevent, or treat illnesses and diseases, such as cancer, diabetes, and more. The goal of these services is to improve and maintain the health of individuals. A city's health should be evaluated from a variety of perspectives: health administration (controlling diseases, preventing epidemics, distributing medical resources, investing in them, etc.), health service organization (hospital, pharmacy, healthcare center, health insurance, etc.), and health status (life expectancy, morbidity, mortality, etc.).

- **Safety**

  From the beginning of civil administration, security and safety have been a fundamental civil service. Nowadays, security concerns are primarily focused on man-made threats, such as crime and terrorism. Safety refers to the actions taken in response to natural disasters and accidents. In both of these areas, ICTs play a crucial role. ICTs allow for greater data collection, analysis, and communication, which are all essential components in responding to man-made and natural threats. They also enable greater collaboration between different government agencies and the public, which is critical in preventing, responding to, and recovering from disasters.

- **Housing**

    The average living space and expenditure in sustainable smart cities are included in this category. As part of the UN's Sustainable Development Goal (SDG) of poverty eradication, it is also very important to reduce slums at the city level. Smart cities provide access to basic services like water and sanitation, as well as provide secure and affordable housing, and also support job creation and income generation. This will help to reduce poverty and create a better living environment for citizens.

- **Culture**

    Library, theater, museum, gallery, and other cultural/knowledge infrastructure are included in sustainable smart city assessments. These places are invaluable to the development of a city, as they provide access to knowledge, history, and culture. They also provide opportunities for education, recreation, and social interaction which help to foster a sense of community and civic engagement.

- **Social Inclusion**

    Toward a sustainable smart city, equity and inclusion should be examined in the following areas: income/consumption equity, access to services and infrastructure for all, openness and public participation, and governance. Social development is strongly influenced by governance and public service. A modern government should be highly efficient and open. Otherwise, stability and development may be compromised by frequent turbulence. Furthermore, ICT will be evaluated in this category in terms of its impact on social harmony and administrative efficiency.

1.8 Smart Cities Around the World

Sustainable cities are not just better for the planet when it comes to their environment. It also has many economic benefits, such as lower energy bills, increased tourism, and new jobs. Sustainable cities also help reduce poverty in the long term, as they are more efficient and cost-effective. In other words, sustainable cities benefit their citizens and are designed to be resilient and adaptable to future challenges. Sustainable cities are designed with convenience in mind and have features such as energy-efficient buildings, green spaces, walkable streets, and bike-friendly paths. These features not only reduce the city's environmental impact, but they also make the city a more pleasant and enjoyable place to live. In order to enhance the sustainability and liveability of real-world smart cities, smart city deployment aims to enhance the competitiveness of cities. People, organizations, and society are experiencing enhanced experiences around the world. The following are a few examples of smart cities around the world.

- **Amsterdam, the Netherlands**

A number of projects were launched in 2006 in Amsterdam, the Netherlands, including connected public lighting within the smart city. In cities, artificial lighting is not only crucial during the daytime, but it is also beneficial for the prestige of the city. This determines the city's ability to attract business and tourists. Due to this, LED lighting was used with smart controllers to reduce consumption. This can result in energy savings of up to 80% and savings of approximately 130 billion euros. In addition, it provides a more general sense of safety and visibility for citizens. Moreover, these systems have been interconnected by incorporating controls via the internet, which leads to more energy savings. Therefore, instead of physical inspections as is done in traditional lighting operations, lighting failures are automatically reported by remote monitoring. Plus, energy consumption was estimated in the past, but smart meters now correctly calculate energy consumption. Moreover, lights are dimmed during low traffic hours to save energy or enhanced when needed to improve safety [65, 66].

- **Chicago and New York, USA**

In Chicago City and Chicago's South Side neighborhoods, residents, groups, the police, and the public use digital tools to share information and reduce violent crime. In this project which is supported by the University of Chicago, by spreading trust among neighbors and local organizations, a general conversation about crime is initiated. To this end, a different source of data collected by police, users, and agencies will be available for a smartphone application that uses a simple mapping interface and GPS. In New York, a platform called City 24/7 has been created to inform, protect, and revitalize the city. To this end, public communication becomes more accessible anytime, anywhere on any device. This is done by integrating information from open government programs, local citizens, and businesses to create awareness. Therefore, this program delivers the information that people need to know. Smart screens are located at bus stops, train stations, shopping malls, and sports facilities to display these data. This can be accessed via Wi-Fi on nearby smartphones or laptops. With these smart screens, people receive information relevant to their immediate proximity, as well as protection by giving local police and fire departments citywide sensing and communication capabilities. With the growth of smart screens all over the city, the amount of data grows and delivers more value to cities, businesses, and citizens.

- **Busan, South Korea**

The government of Busan, South Korea, recognized the potential of ICT to create employment opportunities and economic growth. As a result of Busan's high-quality communication infrastructure, the government was able to expand its cloud infrastructure [67]. It connects universities, industries, citizens, and governments to promote sustainable urban development. As of today, the Busan government, Busan Mobile Application Center (BMAC), and universities are all connected through the cloud. The suggestion made by BMAC regarding the components of a comprehensive workspace, including physical spaces, technology platforms, consulting services, data access and support tools. Moreover, BMAC advocates for the implementation of an application programming interface (API) that enables convenient access to municipal data, including geographical information and intelligent transportation systems. It helps to improve city operations, quality of life, and citizens' access to services. Subsequently, a tremendous number of applications ideas and development proposals have been received, leading to the registration of numerous new businesses as start-ups and individuals registered as professional application developers. As a result, developers through a shared platform can work with the city to create smart city services.

- **Nice, France**

Nice, France, investigated the potential for implementing the Internet of Energy (IoE), testing and validating the IP-enabled technology architecture, economic model, and social benefits of the IoE. As a result, four smart city services were established, including smart lighting, smart circulation, smart waste management, and smart environmental monitoring. This project allowed logging data to be applied to a variety of services. The data collected by traffic sensors are used for smart parking and environmental monitoring.

- **Padova, Italy**

The University of Padova, in collaboration with the municipal government of Padova, has launched a project called Padova Smart City, which is an example of private-public cooperation to run a smart city. Municipalities provide the necessary infrastructure and budget, and universities implement the smart city concept as a theoretical party. In this project, wireless nodes are installed on street light poles to collect environmental and public street lighting data using sensors connected to the internet through gateways. A simple yet accurate mechanism is used to check the correct functioning of the street lighting system by measuring light intensity while gathering environmental parameters such as CO level, air temperature and humidity, vibrations, noise, and so on. Despite its simplicity, this pilot project contains a number of devices and technologies that represent most of the critical issues when designing an urban IoT [68].

- **London, UK**

According to CIMI, London is the second smartest city in the world [69]. As London's population grows by one million in the next decade [70], the city is constantly changing. In order to develop innovative applications, the London data store provides open access to public data. In addition, it includes Wi-Fi connectivity on the tube, intelligent road management, and cycle rental schemes. Digital money is also used in London to enhance savings efficiency. A lot of money is invested in technological advancements in London aimed at improving future cities. The smart transportation system of London is highly recognized as one of the most advanced transportation systems in the world.

- **San Francisco, USA**

According to the US and Canada green city index [71], San Francisco is the greenest city in Northern America. San Francisco utilizes technology to improve the operational performance of buildings, extend transportation systems, centralize waste management procedures, and reduce energy consumption. In other words, San Francisco includes smart transportation, smart energy, and smart community as its main components that serve the citizens of San Francisco. Waste management innovations have been identified as the main aspect of a smart community that contributes to the title of "greenest city." The vision of shared, electric, connected, and automated vehicles (SECAV) was promoted by the SF municipal transportation agency to replace single-occupant vehicles. Ultimately, the SECAV vision has resolved the problem of time-consuming and costly transportation within the city and suburbs [72]. In many ways, San Francisco is a smart city that collects a lot of data. Extended data collection consists of data from connected high-occupancy vehicles, smart traffic signals, and connected safety corridors for pedestrians. The San Francisco Municipal Transportation Agency (SFMTA) plays a major role in the smart transportation initiatives of San Francisco's smart city. Municipal waste management is considered to be one of the most sustainable aspects of smart communities. As stated above, smart energy, smart communities, and next-generation payment systems are a few of the services offered to San Francisco citizens.

- **Barcelona, Spain**

Barcelona is one of the best smart cities in the world as well as in Europe. Due to its knowledge intensiveness, tourist attraction, and main port, Barcelona is considered the industrial spotlight in Spain. During the transformation process of Barcelona, ICT was used to enhance accessibility, transparency, and efficiency of business processes and public administration. Barcelona's smart city is built on infrastructure, information, and human capital [73, 74]. Certain districts of Barcelona have adopted smart city standards in economics, green infrastructure, mobility, science and technology, quality of life, and housing. In light of these facts, Barcelona's city infrastructure has been remodelled to facilitate ICT integration [75]. Using the city model, citizens are able to receive better services that improve their quality of life. Barcelona city has a corporate fiber-optic network, Wi-Fi mesh network, multipurpose and multivendor sensor network, and a public Wi-Fi network to meet these needs [76]. As a result of Barcelona's smart city model, public services have been improved, knowledge is more accessible, infrastructure plans have been improved, and there are many job opportunities with a high level of creativity. An extensive network of sensors in Barcelona enhances the generation of real-time data that supports the manipulation of city services in real time. The provision of services to citizens is carried out by various applications, such as smart transportation, smart healthcare, smart communities, and smart energy. The energy-saving upgrade of street lighting has been a successful step toward smart energy management. In multi-story parking lots, occupancy sensors have improved both efficiency and revenue by increasing the number of parking spaces available. The Barcelona smart city is rapidly developing to serve citizens with the most innovative smart solutions in every aspect of their lives [77–81].

- **Santander City, Spain**

The European Commission in the development of Santander City supported the Smart Santander project. This project provides a large framework for conducting smart city and IoT research experiments. In terms of IoT deployment, this is known as the largest testbed [82]. Additionally, it supports the operations and services of a smart city [83]. Santander City provides experimentation facilities for the heterogeneity of IoT devices, scalability, and real-time mobility scenarios due to its use as a service provision framework and experimentation infrastructure. To overcome existing limitations in executing emerging smart city research, Santander City has integrated a variety of properties.

1.9 Open Research Challenges and Research Direction

The demand for smart cities is diverse and complex. It covers efficiency, sustainability, expenditure, communication links, defense, and surveillance. These factors are driven by a variety of factors, including the economy, environment, communities, and policy. Expenses are the most significant factor in building a smart city. It involves design and service costs [83–85]. Construction costs only once, but operational costs are necessary to keep a smart city running. To make a smart city a reality, it is crucial to minimize development cost. Implementing low-cost construction method will make it easier for cities to function in the long run while putting less strain on their budgets. Efficient operations play a vital role in smart cities as they directly impact service costs. Additionally, promoting sustainability and reducing carbon emissions and urban waste are essential for increasing efficiency and reducing operational expenses. Smart cities must also address pollution and prioritize long term sustainability while improving overall operations and resilience against hazards. These hazards can arise from environmental factor or other system-related issues, including natural disaster. It is important to establish swift recovery mechanism to bounce back from such situations promptly.

Security and privacy have also become necessary to ensure the sustainability of the ubiquitous city and support the IoT, which is the driving force behind future smart cities. The security and privacy challenges posed by the deployment of IoT in smart cities are still an open research challenge. Particular attention should be given to IoT devices, infrastructures, networking, and protocols of the IoT. In order to create a smart city, it is necessary to consider information security and infrastructure. Safety is the most important factor when it comes to building smart cities. Citizens' safety of residents is paramount, and this can also boost design and budgets. Smart cities are allocating more funds and resources to security, while tech companies are developing solutions with upgraded built-in safeguards against hacking and cybercrime [86–88]. Many developers are exploring methods to merge these encryption techniques into new applications using blockchain.

1.10 Organization of the Book

The remainder of this book is organized as follows.

Chapter 1: A smart city is a digitally connected urban environment that uses data, technology, and analytics to improve the quality of life for its citizens. It utilizes sensors and other digital technologies to enable citizens to access services quickly and efficiently. Smart cities also prioritize sustainable development and innovation. The objectives of a smart city is to use technology and data to improve the efficiency of services such as energy, transportation, and waste management, as well as to improve the quality of life for citizens. Smart cities also aim to reduce carbon emissions and promote sustainability. It is built on the idea of utilizing technology to enhance the lives of citizens and create a more efficient and sustainable urban environment. They are enabled by the use of a variety of data-driven technologies such as the Internet of Things, artificial intelligence, and big data. This data is used to develop more efficient ways of managing the city's resources and improve services for citizens. It effectively explains the concept of smart cities, their objectives, and the technologies involved in their implementation. The purpose of this chapter is to examine the importance and need of sustainability and highlights the use of data-driven technologies in creating efficient and innovative urban environments with examples from around the world that offer examples. Furthermore, a detailed aspect of smart city with its requirements, potential components, and open research challenges with opportunities are discussed.

Chapter 2: The growth of data traffic and the interconnection of digital devices to establish reliable communication have made the internet a potential demand for society. To develop a system that securely connects the internet to real-world space would aid in the advancement of a human-centered society that balances economic progress with the resolution of social issues. This chapter provides a detailed overview of Society 5.0 with its requirements, architecture, and components. Society 5.0 has been extensively explored in detail, including its relation to Industry 4.0/5.0 and the role of Society 5.0 in achieving the Sustainable Development Goals (SDGs) of the United Nations. Several emerging communication and computing technologies such as 5G/6G-Internet of Things, edge computing/cloud computing/fog computing, Internet of Everything, blockchain, and beyond networks have also been explored to fulfill the demands of Society 5.0. The potential application of super-smart cities (Society 5.0) with some practical experience of inhabitants is thoroughly discussed. Finally, we highlighted several open research challenges with opportunities.

Chapter 3: To achieve Sustainable Development Goal (SDG) 11 for making cities inclusive, safe, resilient, and sustainable, smart cities are expected to improve the efficiency and effectiveness of urban management, including public services, public security, and environmental protection. In the development of smart cities, enabling technologies has been identified as a key enabler. Globally, cities need intelligent systems to cope with limited space and resources. As a result, smart cities emerged mainly as a result of highly innovative ICT industries and markets, and additionally, they have started to use novel solutions taking advantage of the Internet of Things, big data, and cloud computing technologies to establish a profound connection between each component and layer of a city. In this chapter, we have discussed several key technologies including Internet of Things, Internet of Everything, wireless sensor network, computing technologies, 5G, 6G, cloud computing, edge computing, fog computing, quantum communication, artificial intelligence, machine learning, deep learning, big data/data analytics, blockchain, and cyber-physical systems (CPS) that contribute to the creation of a smart city. It also offers advanced solutions for today's cities.

Chapter 4: Various aspects of smart cities can be made more accessible and applicable with the Internet of Things, which is the backbone of smart cities. In this chapter, we will discuss the concept of sustainable smart cities with respect to Internet of Things, along with its different applications, benefits, and advantages. In addition, IoT technologies are introduced and discussed in terms of how they are integrated into and applied to smart cities in different ways. Further, we discuss how smart cities could be used to develop technology in the future, open issues, and research challenges.

Chapter 5: Currently, the Internet of Things is a highly effective and realistic wireless communication system concept. In this chapter, we have explored the vision of sixth-generation (6G)-Internet of Things-enabled communication networks for sustainable smart cities with its intelligent framework. We proposed a framework that consists of real-time sensing, computing, and communication systems. Further, we have emphasized the architecture and requirements of the 6G-IoT network in relation to the increasing demand for the establishment of a sustainable smart city. In addition, we describe emerging technologies for 6G in terms of artificial intelligence/machine learning, spectrum bands, sensing networks, extreme connectivity, new network architectures, security, and trust that can enable the development of 6G-IoT network architectures to ensure the quality of service (QoS)/quality of experience (QoE) that is large enough to support a massive number of devices. Moreover, we have highlighted potential challenges and research directions for future 6G-IoT wireless communication research for the sustainable smart cities.

Chapter 6: Artificial intelligence is a technology that mimics human intelligence, which is growing in popularity these days. Artificial intelligence is used in many industries today, including marketing, banking, healthcare, and security; robotics; transportation; chatbots; and automated creativity. The use of artificial intelligence has recently become increasingly prevalent in smart city operations. It includes expert systems, natural language processing, speech recognition, and machine vision. Information and communication technologies boost the economy, improve citizens' lives, and assist the government in its duties. It plays a significant role in making cities safer and easier to live in. It also provide people more control over their own homes, monitors traffic, and manages waste. A smart city's sustainable development is investigated in this chapter through the application of artificial intelligence. These applications are used in smart transport, smart healthcare, smart homes/living, smart industries, smart energy, smart agriculture, smart governance, and smart education. Furthermore, existing daily lives and social media examples of artificial intelligence in smart cities are discussed in detail.

Chapter 7: In recent years, smart cities have emerged with energy conservation systems for managing energy in cities as well as buildings. With the Internet of Things, smart cities benefit from a number of smart and extensive applications. As urbanization has grown rapidly in recent years, smart cities must develop effective and sustainable solutions. Further, IoT devices are becoming more energy-efficient, consuming more energy, as well as requiring more energy. In the smart cities, energy management is considered as an interpretivist approach. In this regard, smart city solutions must take into account how energy must be used while also dealing with the issues related to it. Throughout this chapter, we provide a view of energy management and issues facing smart cities. Further, we presented a comprehensive classification of energy management for IoE-based smart cities. Additionally, we discussed energy harvesting in sustainable smart cities, which extend the life of low-power devices, as well as the challenges that come with it.

Chapter 8: In large urban areas and in modern life, transport systems play a vital role. They enable the mobility of not only vehicles but also people residing in cities. The issue of mobility in urban environments is vitally relevant to them. In the Internet of Things, vehicles become intelligent moving nodes or objects within the sensor network, combining the mobile internet and the Internet of Vehicles (IoV). With a focus on deployment in smart cities, this chapter examines the current and emerging paradigms and communication models of IoV. Many authors have surveyed IoV applications, including driver safety, traffic efficiency, and infotainment. The purpose of this chapter is to assess the ability of the IoV to meet the needs of smart cities in terms of data collection, processing, and storage on a large scale. Additionally, the chapter discusses smart cities in relation of the Internet of Vehicles.

Chapter 9: A smart healthcare concept has gradually emerged with the development of information technology. The traditional medical system is being transformed in a broader sense by incorporating cutting-edge information technologies, such as the Internet of Things, big data, cloud computing, and artificial intelligence. This transformation aims to provide more efficient, convenient, and personalized healthcare. In this chapter, we discuss the concept of intelligent healthcare. We list the key technologies that support smart healthcare and discuss the current status of smart healthcare across various fields. Furthermore, we addresss the challenges faced by smart healthcare and discuss potential solutions. In conclusion, we examine the future prospects of smart healthcare and evaluate its potential impact.

Chapter 10: An unmanned aerial vehicle (UAV) is a type of aircraft that does not have a human pilot on board and is capable of controlled flight. UAVs serves various purposes such as monitoring air quality, infrastructure inspection, and offers emergency services, among other applications. These UAVs have proven to a valuable assets in the context of smart cities, offering a range of potential applications supported by key technologies. One notable application is the delivery of goods, which not only reduces traffic congestion but also minimizes emissions compared to the traditional delivery vehicles. It is used to monitor and regulate urban development, transportation, and infrastructure. These applications offers real-time data to cities for better decision-making and resource allocation. This contributes to create more sustainable, efficient, and equitable smart cities. Our goal in this chapter is to delve into how UAVs address challenges and contribute to the sustainable development of smart cities. Finally, we evaluate UAVs prospects and outline future research opportunities in this rapidly evolving field.

1.11 Conclusion

Smart cities should be designed with sustainability in mind, taking into account the environmental impact of their construction and operations. The concept of a sustainable smart city holds immense potential for addressing urban challenges and improving the quality of life for its residents. Sustainable city planning should also prioritize the needs of people, ensuring that everyone has access to basic services, safe housing, and green spaces. Technology is a powerful tool in making cities more sustainable, helping to reduce resource use and emissions. Smart technologies are used to optimize city operations, provide better services to citizens, and improve the quality of life in urban areas. Through the integration of advanced technologies such as the IoT, data analytics, and renewable energy sources, sustainable smart cities achieve energy efficiency, resource optimization, and enhanced connectivity. Sustainable smart cities must include components such as energy and water management, waste disposal, transportation, public safety, and other elements that can be managed efficiently through the use of technology. These cities prioritize environmental conservation, social equity, and economic prosperity, establishing a harmonious balance between people, the environment, and technological advancements. However, the development of sustainable smart cities necessitates careful planning, collaboration between stakeholders, robust security measures, and a focus on privacy to ensure long-term success. By embracing sustainability as a guiding principle, smart cities serve as catalysts for positive change, fostering innovation, resilience, and a better future for urban communities worldwide.

References

1. M.M.P. Rana, Urbanization and sustainability: Challenges and strategies for sustainable urban development. Bangladesh Environ. Dev. Sustain. **13**(1), 237–256 (2011)
2. F. Cugurullo, Exposing smart cities and eco-cities: Frankenstein urbanism and the sustainability challenges of the experimental city. Environ. Plann. A: Econ. Space **50**(1), 73–92 (2018)

3. F. Montori, L. Bedogni, L. Bononi, A collaborative internet of things architecture for smart cities and environmental monitoring. IEEE Internet Things J. **5**(2), 592–605 (2018)
4. W. Li, H. Song, F. Zeng, Policy-based secure and trustworthy sensing for internet of things in smart cities. IEEE Internet Things J. **5**(2), 716–723 (2018)
5. P.M. Santos, C. Queiros, S. Sargento, A. Aguiar, J. Barros, J.G.P. Rodrigues, S.B. Cruz, T. Lourenco, P.M. d'Orey, Y. Luis, C. Rocha, S. Sousa, S. Crisostomo, PortoLivingLab: An IoT-based sensing platform for smart cities. IEEE Internet Things J. **5**(2), 523–532 (2018)
6. C. Harrison, B. Eckman, R. Hamilton, P. Hartswick, J. Kalagnanam, J. Paraszczak, P. Williams, Foundations for smarter cities. IBM J. Res. Dev. **54**(4), 1–16 (2010)
7. B.N. Silva, M. Khan, K. Han, Towards sustainable smart cities: A review of trends, architectures, components, and open challenges in smart cities. Sustain. Cities Soc. **38**, 697–713 (2018)
8. A. Deshpande, C. Guestrin, S.R. Madden, J.M. Hellerstein, W. Hong, Model driven data acquisition in sensor networks, in *Proceedings of the 13th International Conference on Very Large Data Bases*, vol. 30(1), (2004), pp. 588–599
9. S. Al-Nasrawi, C. Adams, A. El-Zaart, A conceptual multidimensional model for assessing smart sustainable cities. J. Inf. Syst. Technol. Manag. **12**, 541–558 (2015)
10. M. Ibrahim, C. Adams, A. El-Zaart, Paving the way to smart sustainable cities: Transformation models and challenges. JISTEM J. Inf. Syst. Technol. Manag. **12**, 559–576 (2015)
11. M. Hojer, J. Wangel, Smart sustainable cities: Definition and challenges, in *ICT Innovations for Sustainability*, (Springer, Cham, 2015), pp. 333–349
12. F. Mosannenzadeh, D. Vettorato, Defining smart city. A conceptual framework based on keyword analysis. TeMA J. Land Use Mobil. Environ. (2014). https://doi.org/10.6092/1970-9870/2523
13. V. Albino, U. Berardi, R.M. Dangelico, Smart cities: Definitions, dimensions, performance, and initiatives. J. Urban Technol. **22**(1), 3–21 (2015)
14. A. Caragliu, C. Del Bo, P. Nijkamp, Smart innovative cities: The impact of Smart City policies on urban innovation. Technol. Forecast. Soc. Chang. **142**, 373–383 (2019)
15. IDA Singapore. iN2015 Masterplan (2012). Available online: http://www.ida.gov.sg/~/media/Files/Infocomm%20Landscape/iN2015/Reports/realisingthevisionin2015.pdf. Accessed on 14 Mar 2023
16. A.M. Hassan, H. Lee, The paradox of the sustainable city: Definitions and examples. Environ. Dev. Sustain. **17**, 1267–1285 (2015)
17. N. Duxbury, C. Cullen, J. Pascual, "Cities, culture and sustainable development", Cultural policy and governance in a new metropolitan age. Cult. Glob. Ser. **5**, 73–86 (2012)
18. A. Kramers, M. Hojer, J. Wangel, Planning for smart sustainable cities: Decisions in the planning process and actor networks, in *Proceedings of the 2nd International Conference on ICT for Sustainability (ICTS)*, (Stockholm, Sweden, 24–27 August 2014), pp. 299–305
19. S.E. Bibri, J. Krogstie, Smart sustainable cities of the future: An extensive interdisciplinary literature review. Sustain. Cities Soc. **31**, 183–212 (2017)
20. T. Nam, T.A. Pardo, Conceptualizing smart city with dimensions of technology, people, and institutions, in *Proceedings of the 12th Annual International Digital Government Research Conference: Digital Government Innovation in Challenging Times*, (College Park, MD, USA, 12–15 June 2011), pp. 282–291
21. R. Giffinger, H. Gudrun, Smart cities ranking: An effective instrument for the positioning of the cities? ACE Archit. City Environ. **4**, 7–26 (2010)
22. L. Anavitarte, B. Tratz-Ryan, *Market insight:'smart cities' in emerging markets*, vol 12 (Gartner, Stamford, 2010), pp. 39–61
23. S. Lombardi, H.F. Giordano, W. Yousef, Modelling the Smart City performance. Innov. Eur. J. Soc. Sci. Res. **25**, 137–149 (2012)
24. R.W.S. Ruhlandt, The governance of smart cities: A systematic literature review. Cities **81**, 1–23 (2018)
25. J. Curzon, A. Almehmadi, K. El-Khatib, A survey of privacy enhancing technologies for smart cities. Pervasive Mob. Comput. **55**, 76–95 (2019)

References

26. A. Elsaeidy, K.S. Munasinghe, D. Sharma, A. Jamalipour, Intrusion detection in smart cities using restricted Boltzmann machines. J. Netw. Comput. Appl. **135**, 76–83 (2019)
27. D. Hill, Smarts and the City. Waste Manag. Environ. **21**, 20 (2010)
28. S.A. Remaco, *Jessica–Joint European Support for Sustainable Investment in City Areas-Jessica Instruments for Energy Efficiency in Greece–Evaluation Study* (Final Report, European Commission, Brussels, 2010)
29. B. Cohen, *What Exactly Is a Smart City?* (Fast CoExist, 2012)
30. https://highlights.hitachi.com/_tags/Smart_City. Accessed on 21 Mar 2023
31. J.S. Hwang, O.G. Young, H. Choe, Smart cities seoul: A case study, in *ITU-T Technology Watch Report*, (ITU, Geneva, 2013)
32. A. Ramamurthy, M.D. Devadas, Smart sustainable cities: An integrated planning approach towards sustainable urban energy systems, India. Int. J. Hum. Soc. Sci. **7**(1), 252–272 (2013)
33. C. Manville, G. Cochrane, C.A.V.E. Jonathan, J. Millard, J.K. Pederson, R.K. Thaarup, A.L.I.E.B.E. WiK, M.W. Wik, *Mapping smart cities in the EU* (Publications Office, Luxembourg, 2014)
34. BSI. Smart Cities—Vocabulary, *BSI Standards Publication, PAS 180:2014* (British Standards Institute (BSI), London, 2014)
35. International Telecommunications Union (ITU). Technical Specifications on "Setting the Framework for an ICT Architecture of a Smart Sustainable City" (SSC-0345) (2015). Available online: http://www.itu.int/en/ITUT/focusgroups/ssc/Pages/default.aspx. Accessed on 8 Feb 2023
36. S.P. Mohanty, U. Choppali, E. Kougianos, Everything you wanted to know about smart cities: The internet of things is the backbone. IEEE Consumer Electron. Mag. **5**, 60–70 (2016)
37. J. An, F. Le Gall, J. Kim, J. Yun, J. Hwang, M. Bauer, M. Zhao, J. Song, Toward global IoT-enabled smart cities interworking using adaptive semantic adapter. IEEE Internet Things J. **6**(3), 5753–5765 (2019)
38. R. Diaz-Diaz, D. Perez-Gonzalez, Implementation of social media concepts for e-government: Case study of a social media tool for value cocreation and citizen participation. J. Org. End User Comput. **28**(3), 104–121 (2016)
39. H. Truong, S. Dustdar, A survey on cloud-based sustainability governance systems. Int. J. Web Inf. Syst. **8**(3), 278–295 (2012)
40. I. Hwang, C.E. Seah, Intent-based probabilistic conflict detection for the next generation air transportation system. Proc. IEEE **96**(12), 2040–2059 (2008)
41. L. Foschini, T. Taleb, A. Corradi, D. Bottazzi, M2M-based metropolitan platform for IMS-enabled road traffic management in IoT. IEEE Commun. Mag. **49**(11), 50–57 (2011)
42. Y. Wang, S. Ram, F. Currim, E. Dantas, L.A. Saboia, A big data approach for smart transportation management on bus network, in *Proceedings of the IEEE International Smart Cities Conference*, ((ISC2), Trento, Italy, September 2016), pp. 1–6
43. M. Mazzarello, E. Ottaviani, A traffic management system for real-time traffic optimisation in railways. Transp. Res. B Methodol. **41**(2), 246–274 (2007)
44. F. Corman, A. D'Ariano, D. Pacciarelli, M. Pranzo, Bi-objective conflict detection and resolution in railway traffic management. Trans. Res. C Emerg. Technol. **20**(1), 79–94 (2012)
45. F. Montori, L. Bedogni, M. Di Felice, L. Bononi, Machine-to-machine wireless communication technologies for the Internet of Things: Taxonomy, comparison and open issues. Pervasive Mob. Comput. **50**, 56–81 (2018)
46. M. Khan, B.N. Silva, K. Han, Internet of things-based energy aware smart home control system. IEEE Access **4**, 7556–7566 (2016)
47. J. Han, C.S. Choi, W.K. Park, I. Lee, S.H. Kim, Smart home energy management system including renewable energy based on ZigBee and PLC. IEEE Trans. Consum. Electron. **60**, 198–202 (2014)
48. E. Shirazi, S. Jadid, Autonomous self-healing in smart distribution grids using multi agent systems. IEEE Trans. Ind. Inform. **15**(12), 6291–6301 (2018)

49. L. Catarinucci, D. De Donno, L. Mainetti, L. Palano, L. Patrono, M.L. Stefanizzi, et al., An IoT-aware architecture for smart healthcare systems. IEEE Internet Things J. **2**, 515–526 (2015)
50. H. Zhang, J. Li, B. Wen, Y. Xun, J. Liu, Connecting intelligent things in smart hospitals using NB-IoT. IEEE Internet Things J. **5**(3), 1550–1560 (2018)
51. B.R. Haverkort, A. Zimmermann, Smart industry: How ICT will change the game! IEEE Internet Comput. **21**, 8–10 (2017)
52. F. Tao, J. Cheng, Q. Qi, IIHub: An industrial internet-of-things hub toward smart manufacturing based on cyber-physical system. IEEE Trans. Ind. Inform. **14**, 2271–2280 (2018)
53. P. Trakadas, P. Simoens, P. Gkonis, L. Sarakis, A. Angelopoulos, A.P. Ramallo-Gonzalez, A. Skarmeta, C. Trochoutsos, D. Calvo, T. Pariente, et al., An artificial intelligence-based collaboration approach in industrial IoT manufacturing: Key concepts, architectural extensions and potential applications. Sensors **20**(19), 5480/1–20 (2020)
54. J. Wan, J. Yang, Z. Wang, Q. Hua, Artificial intelligence for cloud-assisted smart factory. IEEE Access **6**, 55419–55430 (2018)
55. J.R. Gil-Garcia, T.A. Pardo, T. Nam, What makes a city smart? Identifying core components and proposing an integrative and comprehensive conceptualization. Inf. Polity **20**(1), 61–87 (2015)
56. M. De Waal, M. Dignum, The citizen in the smart city. How the smart city could transform citizenship. It-Inf. Technol. **59**(6), 263–273 (2017)
57. P. Neirotti, A. De Marco, A.C. Cagliano, G. Mangano, F. Scorrano, Current trends in smart city initiatives: Some stylized facts. Cities **38**, 25–36 (2014)
58. M. Sharholy, K. Ahmad, G. Mahmood, R. Trivedi, Municipal solid waste management in Indian cities – A review. Waste Manag. **28**, 459–467 (2008)
59. A. Koubaa, A. Aldawood, B. Saeed, A. Hadid, M. Ahmed, A. Saad, H. Alkhouja, A. Ammar, M. Alkanhal, Smart palm: An IoT framework for red palm weevil early detection. Agronomy **10**(7), 987/1–21 (2020)
60. M.J. O'Grady, D. Langton, G.M.P. O'Hare, Edge computing: A tractable model for smart agriculture? Artif. Intell. Agric. **3**, 42–51 (2019)
61. http://www.intel.ie/content/dam/www/public/us/en/documents/flyers/education-ict-bene fits-infographic.pdf. Accessed on 01 Mar 2023
62. A. Quijano, J.L. Hernandez, P. Nouaille, M. Virtanen, B. Sanchez-Sarachu, F. Pardo-Bosch, J. Knieilng, Towards sustainable and smart cities: Replicable and KPI-driven evaluation framework. Buildings **12**(2), 233/1–12 (2022)
63. M. Hara, T. Nagao, S. Hannoe, J. Nakamura, New key performance indicators for a smart sustainable city. Sustainability **8**(3), 206/1–19 (2016)
64. Z. Sang, K. Li, ITU-T standardization activities on smart sustainable cities. IET Smart Cities **1**(1), 3–9 (2019)
65. Amsterdam Smart City. Available online: https://amsterdamsmartcity.com/. Accessed on 12 Mar 2023
66. The Internet of Everything for Cities. Available online: http://pie.pascalobservatory.org/sites/default/files/ioe-smart-city_pov.pdf. Accessed on 24 Mar 2023
67. E. Strickland, Cisco bets on South Korean smart city. IEEE Spectr. **48**(8), 11–12 (2011)
68. A. Zanella, N. Bui, A. Castellani, L. Vangelista, M. Zorzi, Internet of things for smart cities. IEEE Internet Things J. **1**(1), 22–32 (2014)
69. P.R. Berrone, J. Enric, *IESE Cities in Motion Index* (IESE Business School University of Navarra, 2016). Accessed on 8 Mar 2023
70. S. L. Board, *Smart London Plan*, (2016). Accessed on 14 Mar 2023
71. S. F. Environment, Designing a smarter, more Sustainable: San Francisco, in *San Francisco Department of the Environment*, (2016). Accessed on 25 Jan 2023
72. "City of San Francisco: Meeting the Smart City Challenge, San Francisco Smart City Challenge", Municipal Transportation Agency, S. F. M. T. A. (2016). Accessed on 18 Jan 2023
73. T. Bakıcı, E. Almirall, J. Wareham, A smart city initiative: The case of Barcelona. J. Knowl. Econ. **4**, 135–148 (2013)

References

74. N. Leon, Attract and connect: The 22@Barcelona innovation district and the internationalization of Barcelona business. Innov. Org. Manag. **10**, 235–246 (2014)
75. Smart City | Servei de Premsa | El Web de la Ciutat de Barcelona. Available online: http://ajuntament.barcelona.cat/premsa/tag/smart-city/. Accessed on 8 Jan 2023
76. Laursen, L, *City Saves Money, Attracts Businesses with Smart City Strategy*. Available online: https://www.technologyreview.com/s/532511/barcelonas-smart-city-ecosystem/. Accessed on 8 Jan 2023
77. TMBAPP (Metro Bus Barcelona) | Apps | iTunes | apps4BCN | All the Apps You Need for Barcelona! Available online: http://apps4bcn.cat/en/app/tmbapp-metro-bus-barcelona/111. Accessed on 8 Jan 2023
78. Transport & Traffic | IOS | The Best App Selection for Barcelona | Apps4bcn | All the Apps You Need for Barcelona! Available online: http://apps4bcn.cat/en/apps/index/Category:transport-i-tr-nsit. Accessed on 6 Jan 2023
79. UrbanStep Barcelona | Apps | iTunes | apps4BCN | All the Apps You Need for Barcelona! Available online: http://apps4bcn.cat/en/app/urbanstep-barcelona/110. Accessed on 9 Jan 2023
80. I. Capdevila, M.I. Zarlenga, Smart city or smart citizens? The Barcelona case. J. Strateg. Manag. **8**(3), 266–282 (2015)
81. P. Sotres, J.R. Santana, L. Sanchez, J. Lanza, L. Munoz, Practical lessons from the deployment and management of a smart city internet-of-things infrastructure: The SmartSantander testbed case. IEEE Access **5**, 14309–14322 (2017)
82. L. Sanchez, L. Munoz, J.A. Galache, P. Sotres, J.R. Santana, V. Gutierrez, R. Ramdhany, A. Gluhak, S. Krco, E. Theodoridis, D. Pfisterer, Smart Santander: IoT experimentation over a smart city testbed. Comput. Netw. **61**, 217–238 (2014)
83. M.A. Ahad, S. Paiva, G. Tripathi, N. Feroz, Enabling technologies and sustainable smart cities. Sustain. Cities Soc. **61**, 102301/1–12 (2020)
84. H.H. Khan, M.N. Malik, R. Zafar, F.A. Goni, A.G. Chofreh, J.J. Klemes, Y. Alotaibi, Challenges for sustainable smart city development: A conceptual framework. Sustain. Dev. **28**(5), 1507–1518 (2020)
85. H. Zahmatkesh, F. Al-Turjman, Fog computing for sustainable smart cities in the IoT era: Caching techniques and enabling technologies-an overview. Sustain. Cities Soc. **59**, 102139/1–15 (2020)
86. H. Kim, H. Choi, H. Kang, J. An, S. Yeom, T. Hong, A systematic review of the smart energy conservation system: From smart homes to sustainable smart cities. Renew. Sust. Energ. Rev. **140**, 110755/1–17 (2021)
87. A. Ahmad, G. Jeon, C.W. Yu, Challenges and emerging technologies for sustainable smart cities. Indoor Built Environ. **30**(5), 581–584 (2021)
88. F. Almalki, S.H. Alsamhi, R. Sahal, J. Hassan, A. Hawbani, N.S. Rajput, A. Saif, J. Morgan, J. Breslin, Green IoT for eco-friendly and sustainable smart cities: Future directions and opportunities. Mob. Netw. Appl. **21**(19), 1–25 (2021)

Chapter 2
Sustainable Smart City to Society 5.0

2.1 Introduction

A sustainable smart city is an urban environment that plays a crucial role in economic and societal development by offering residents convenient access to services anytime and anywhere. These cities are designed to prioritize efficiency and sustainability through the utilization of data and technology. Their aim is to reduce waste, energy consumption, and vehicle emissions while providing services such as healthcare, education, and transportation to meet the needs of their citizens. The United Nations World Urbanization Prospects 2018 predicts that nearly 70% of the world's population will reside in urban areas [1]. Nevertheless, several megacities are already experiencing significant population growth, prompting researchers to focus on sustainable smart cities solutions to enhance their livability and sustainability. It aims to create liveable, sustainable, and prosperous societies through the use of state-of-the-art technologies in waste management, mobility, water, building, heating, cooling, and smart energy systems but also puts strains on the infrastructure of a city [2].

Furthermore, the infrastructure and services of sustainable smart cities should prioritize efficiency, environmental friendliness, and universal accessibility. The sustainable smart city will, therefore, improve the quality of life of residents and ultimately support the United Nations Sustainable Development Goals (UN SDGs). In order to create sustainable smart cities, urban planning and administration must consider social, environmental, and economic aspects. The focus is on finding technological solutions to make urban development more efficient. In most cases, sustainable development refers to issues within a city's administrative boundaries. This means that the city must consider all aspects of the environment and its inhabitants, such as air quality, access to clean water, access to green spaces, transportation, clean energy, and waste management. Technological solutions must be used to address these issues, such as sensors to monitor air quality, renewable energy systems, intelligent transportation systems, and smart waste management systems. These technological solutions help to reduce pollution, conserve energy, and improve the quality of life for the city's inhabitants. As urbanization continues to grow, the demand for efficient management and application of natural resources, rapid and efficient transport and commuting, safety concerns, and the growing desire for a healthy environment with efficient energy consumption are all expected to drive the growth of the sustainable smart city market [3]. Globally, smart cities will focus on smart transportation, smart buildings, smart utilities, smart citizen, smart healthcare, smart education, and smart energy. Over the forecast period, the global smart cities' market size is expected to grow from USD 457.0 billion in 2021 to USD 873.7 billion by 2026, at a compound annual growth rate (CAGR) of 13.8%.

The revitalization of industry will, in turn, have a transformative effect on society. In the last decade, Industry 4.0 has shifted away from fundamental ideas such as social justice and sustainability and focuses more on digitalization and artificial intelligence (AI)-driven technologies. It refers to a vision for technological innovation and further technological development of European industries in the future [4]. It explains how industry will use technology to better manage itself in a changing world and economy. As of today, Industry 5.0 emphasizes the importance of research and innovation in helping industry provide long-term service to humanity. Its purpose is to serve as a platform for developing a vision of the future of European industry that is collaborative and cooperative. Industry 5.0 is an open and evolving concept that involves the integration of the physical and virtual worlds. The concept includes innovations and cost-saving measures that benefit all, whether they are consumers, workers, investors, or the environment.

The three attributes of Industry 5.0 are human centricity, sustainability, and resilience. By adhering to the boundaries of our planet and prioritizing workers' well-being, industrial companies can contribute to prosperity. The industry has the potential to accomplish social goals beyond creating jobs through this approach. It is imperative for industries to be environmentally aware and recycle and repurpose natural resources in a circular manner while minimizing pollution and waste. Efforts must be made to reduce energy consumption and greenhouse gas emissions in order to maintain the sustainability of the environment. This will meet the needs of the current generation without compromising the needs of future generations. By using technologies like AI and additive manufacturing, we can reduce waste and improve resource efficiency. Ultimately, we can say that Industry 5.0 is beneficial for both employers and employees. Since both describe fundamental changes in our economy and society, Industry 5.0 and Society 5.0 are related. Society 5.0, however, aims to achieve a balance between economic growth, management of society, and environmental concerns. Technology integrates physical and virtual spaces, so it is not limited to the manufacturing industry, but also deals with a broader range of social concerns. Hence, Society 5.0 is a society based on advanced technologies and augmented reality, which primarily benefits citizens and facilitates their daily lives [5, 6].

In Society 5.0, the main objective is to achieve a balance between commercial progress and social issues via a network that closely connects the internet and the

2.1 Introduction

real world. Despite its futuristic nature, Society 5.0 appears to be radically different from previous societies. Human progress began in the hunter era of human progress (Society 1.0) and even during the earliest civilizations. In addition, society has grown from an agricultural (Society 2.0) to an industrial (Society 3.0) era and is evolving into an information society (Society 4.0). As shown in Fig. 2.1, AI will affect every aspect of society. The Internet of Things (IoT) has transformed the way big data is collected. A key objective of Society 5.0 is to build a society where societal constraints are met through the use of emerging technologies such as 5th generation/6th generation (5G/6G), IoT, AI, and big data, along with other emerging communication, computing, sensing, and actuation technologies in the industrial and social spheres. Industry 4.0/5.0 has led to a global trend toward a more sophisticated model of society, necessitating the creation of a super-intellectual Society 5.0. In light of the worldwide trend toward a more advanced model of society, which is directly related to the industrial revolutions, it is necessary to establish a highly intellectual Society 5.0. In addition, it is defined as a civilization that is human-centered and focuses on economic growth and social issues through a system tightly linked to the internet and the physical world.

At present, the IoT is one of the key technologies for smart cities, offering solutions to the challenges inherent in traditional urban development. Through wireless or wired connections, a network of physical objects integrates embedded

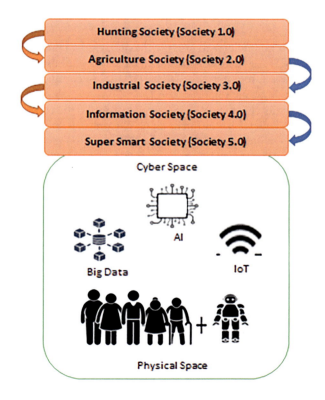

Fig. 2.1 Realization of Society 5.0 with its evolutions and involvement of emerging technologies

technologies for sensing, collecting, computing/processing, communicating, and interacting with their internal states or external environment. It creates new applications and services by using unique addressing schemes and network infrastructures. Data from IoT devices can be stored at remote locations for processing and storage, which provides the ability to collect many relevant data points about individuals in a sustainable smart city. As an alternative way of establishing sustainable human settlements, new urban approaches have recently been developed as Society 5.0 [7].

An IoT-based smart city/Society 5.0 concept has been formulated for realizing a sustainable smart city. Researchers are being urged to focus on developing IoT applications for sustainable smart cities and emphasize the importance of data and information exchange within IoT services, given the significant progress in data collection within the IoT community [8, 9].

Further, AI enables densely populated cities to respond more effectively to urban challenges. By leveraging the power of AI, researchers can develop applications that analyze large datasets in order to identify patterns and trends in the data. This can help cities better understand their needs and make more informed decisions about how to best utilize resources to meet those needs. Additionally, AI-enabled applications help cities better manage traffic and energy usage, which are key components of sustainable smart cities. Moreover, the city's expansion incorporates ICT for efficient resource management, which enhances the quality of life for residents and improves their services. In a smart city infrastructure, ICT can improve a variety of aspects of society, such as health quality, mobility management, and resource management. As explained in [10], it is crucial to connect social, physical, business, and ICT infrastructures in order to maximize urban intelligence. Furthermore, an exploration is conducted on the living standards of a sustainable smart city's and an environmental monitoring system [11, 12]. According to [13], the integration of ICT technology can improve cities.

In this chapter, we have given priority to the concept of Society 5.0 in conjunction with sustainable smart cities. We present an economic, societal, and technical analysis of the sustainable smart city concept. Further, we highlight the involvement of modern technologies in smart city design. There is a wide range of literature that describes the sustainable smart city concept from different perspectives, including AI, blockchain, big data, IoT, and 5G. This chapter provides an overview of the fundamental challenges in the development of a super-intelligent smart city, making a significant contribution to the field. The main key points covered are as follows:

- In order to gain a better understanding of Society 5.0, we have reviewed various aspects of it in the available literature. A number of additional features of Society 5.0 were discussed in comparison with the Industry 4.0/5.0 evolutions.
- We have explored and connected the United Nations Sustainable Development Goals (UN SDGs) with Society 5.0.
- Society 5.0 has a number of issues and challenges despite extensive research and development. Lastly, we describe these challenges in terms of essential social needs, privacy, security, and rights. We also explore potential research directions to realize the vision of Society 5.0.

The rest of the chapter is organized as follows. A background of sustainable smart cities toward the Industrial Revolution and requirements of Society 5.0 are presented in Sects. 2.2 and 2.3, respectively. Sustainable development goals for Society 5.0 are presented in Sect. 2.4. The potential research challenges, issues, and opportunities are explored in Sect. 2.5. Finally, Sect. 2.6 concludes the chapter and recommends the future scope.

2.2 Sustainable Smart Cities Toward the Industrial Revolution

A new wave of digital innovation is sweeping the world in the Fourth Industrial Revolution. Numerous technological advancements are occurring in our everyday lives, ranging from autonomous vehicles to intelligent robotics. Smart cities are closely associated with the concept of the new industrial revolution. It is possible to think of smart cities as the realization of the Fourth Industrial Revolution in a spatial context. Globally, a number of cities are transforming themselves into smart cities through urban regeneration. This chapter recognizes these changes and seeks to understand how our cities are affected by the digital revolution.

2.2.1 Industry 4.0/5.0

During the First Industrial Revolution, mechanical power was generated from different sources, followed by the use of electricity for assembly lines in 1780. The production industry has automated activities using information technology. The Fourth Industrial Revolution employed IoT and the cloud to connect virtual and physical space [14, 15]. Thus, industry pioneers anticipate a future where human intelligence and machines are integrated for a better solution. Through the integration of business processes and production systems, the Fourth Industrial Revolution transformed manufacturing agents into cyber-physical systems (CPS). From suppliers to production lines to end users, all components of the manufacturing industry's supply chain must be integrated using IoT [16].

In Industry 4.0, CPS is used to communicate with all entities connected to the IoT. Therefore, a significant amount of data is processed in a cloud environment. The underlying concept of Industry 4.0 is smart manufacturing, which incorporates many technologies, including cloud computing, IoT, AI, robotics, big data analytics, ambient intelligence, virtual reality, and cyber security [17, 18]. A major advantage of Industry 4.0 is the reduction of production, logistics, and quality management costs. Due to this, Industry 4.0 will be pushed backward, and labor unions will raise resistance, thereby preventing its full adoption [19]. Due to this, the next industrial revolution has been driven by a technological solution that provides pollution-free manufacturing processes [5, 20].

The Fifth Industrial Revolution has focused on intelligent manufacturing by bringing back human intelligence to the production floor by enabling the robots to share and collaborate with humans. As shown in Table 2.1, the authors provide literature-based definitions of the Fifth Industrial Revolution. The Industry 5.0 has brought back the humans to co-work with machines (robots) on factory floors, thus utilizing human intelligence and creativity for intelligent processes [27]. Humans will share and collaborate with the COBOTs without fear of job insecurity, thus resulting in value-added services. As Industry 5.0 is yet to evolve to the fullest, various industry practitioners and researchers have provided various definitions.

2.2.2 Society 5.0

Using technology, Society 5.0 aims to increase the growth rate of Japanese society and reduce its inverse impacts, such as an aging population [28–30]. Realization of Society 5.0 with its evolutions and involvement of emerging technologies is depicted

Table 2.1 Definitions of the Fifth Industrial Revolution

Definition 1. "Industry 5.0 is a first industrial evolution led by the human based on the 6R (Recognize, Reconsider, Realize, Reduce, Reuse and Recycle) principles of industrial upcycling, a systematic waste prevention technique and logistics efficiency design to valuate life standard, innovative creations and produce high-quality custom products [21] by Michael Rada, founder and Leader, Industry 5.0"
Definition 2. "Industry 5.0 brings back the human workforce to the factory, where human and machine are paired to increase the process efficiency by utilizing the human brainpower and creativity through the integration of workflows with intelligent systems [22]"
Definition 3. "European Economic and Social committee states that the new revolutionary wave, Industry 5.0, integrates the swerving strengths of cyber–physical production systems (CPPS) and human intelligence to create synergetic factories [23]. Furthermore, to address the manpower weakening by Industry 4.0, the policymakers are looking for innovative, ethical and human-centered design"
Definition 4. "Friedman and Hendry suggest that Industry 5.0 compels the various industry practitioners, information technologists and philosophers to focus on the consideration of human factors with the technologies in the industrial systems [24]"
Definition 5. "Industry 5.0 is the age of Social Smart factory where cobots communicate with the humans [25]. Social Smart Factory uses enterprise social networks for enabling seamless communication between human and CPPS components"
Definition 6. "Industry 5.0, a symmetrical innovation and the next generation global governance, is an incremental advancement of Industry 4.0 (asymmetrical innovation). It aims to design orthogonal safe exits by segregating the hyperconnected automation systems for manufacturing and production [26]"
Definition 7. "Industry 5.0 is a human-centric design solution where the ideal human companion and cobots collaborate with human resources to enable personalize autonomous manufacturing through enterprise social networks. This, in turn, enables human and machine to work hand in hand. Cobots are not programmable machines, but they can sense and understand the human presence. In this context, the cobots will be used for repetitive tasks and labor-intensive work, whereas human will take care of customization and critical thinking (thinking out of the box)"

in Fig. 2.1. The utilization of technologies such as CPS, AI, and the IoT has recently led to the establishment of a number of efficient and effective systems [31, 32]. Technology is resulting in a large amount of data being collected in both business and everyday lives, and all of these data are referred to as big data. In the business world, the precision errors made by AI on big data have a significant impact on the business because the data is unprocessed. In order to store and analyze large amounts of data, a cloud system is required. Several companies with internet connectivity offer cloud-based systems for hire.

According to the UN SDGs, Japan has agreed to establish Society 5.0 to address various societal issues. Despite this, it is believed that AI will solve many problems associated with Society 5.0, such as unemployment, poverty, and air pollution. There are two prominent types of AI: machine learning (ML) and deep learning (DL). With ML, algorithms become smarter and smarter over time without human intervention, and they self-learn. In other words, it involves enabling a machine to learn more efficiently and intelligently. To provide intelligently effective results, ML enables AI solutions to be faster and smarter. As AI technology continues to influence society and modern business practices, such as strategy, logistics, management, and manufacturing, ML approaches will become increasingly significant. Currently, it is used for a wide range of tasks, especially for applications that require data collection, analysis, and response. In healthcare, retail, manufacturing, banking and finance, and transportation, this technique is highly effective. Another subset of AI, called DL approaches, is also gaining popularity in a variety of social settings. However, there are many challenges, including data security, management, and scalability. According to [33], DL-based IoT-oriented architecture for a secure society is advantageous for addressing these issues. To address communication latency, centralization, and scalability, a DL-based cloud is used at the application layer. As a result, Society 5.0 applications are able to use cost-effective, high-performance computing resources [34].

In addition, Society 5.0 addresses numerous social issues, including healthcare, hunger, wealth, food and clean water, modern agricultural methods, and women's rights [35]. Even so, the Society 5.0 incorporates all the SDGs. In order to develop infrastructure systems effectively, Society 5.0 consists of a plan for infrastructure systems.

The goal of a system is to implement enabling technologies to solve problems. This will play a significant role in moving toward Society 5.0 [36]. As part of Society 5.0, women will be better integrated into the workplace, and smart technology will be used to develop agricultural and food applications [37]. In many industries, CPS have a significant impact on economies and society. A wide range of applications may be possible, from agriculture to manufacturing and from critical infrastructure to assistive living [38]. In the concept of "Society 5.0," innovation and population well-being are emphasized. Social issues such as aging populations, labor shortages, and slower economic growth will influence the economy and society. Its primary objective is to create a human-centered society through the use of technology and creativity. The Japanese government aims to become the first

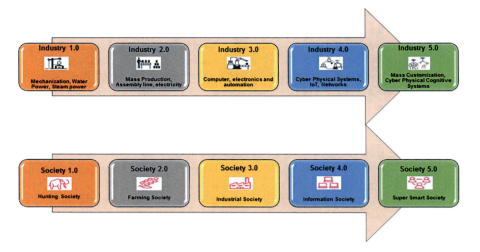

Fig. 2.2 Industrial Revolution to Society 5.0

country to demonstrate how to grow through innovation despite a declining population [39].

While Industry 4.0 focuses only on industrial output, Society 5.0 seeks to encompass all aspects of society [40]. Figure 2.2 shows the evolution of industries to Society 5.0. Industry 4.0 involves establishing systems between private enterprises and academia, with government funding provided for research and development [41]. Industry 4.0/5.0 and Society 5.0 use similar technologies, including AI, CPS, IoT, big data, robotics, and the cloud [34]. As Industry 4.0/5.0 is efficient in a confined region because it is specific to an industry, Society 5.0 is efficient in a broader area since it involves society as a whole. An analysis of 51 OGD (open government data) portals is presented in [42]. By examining relevant OGD portals, a review of 60 countries and their OGD portals was conducted to determine if they meet the Society 5.0 and Industry 4.0 trends. To be a leader, the industry must transform itself on its own [43]. As discussed in [22], open data is related to Industry 4.0 and Society 5.0. Human-centered action, sustainable development, and the physical to digital to physical loop in conjunction with OGD portal, combined with geospatial data and real-time data, contribute to the development of a human-centered society through a sustainable development model.

To achieve a sustainable and inclusive society, humans must be involved in the process of developing policy and making decisions. In [19], the discussion revolves around how AI can be utilized to develop evidence-based policies that contribute to the achievement of the SDGs in Society 5.0. In a study conducted by [44], 60 government portals were compared against the expectations of Society 5.0. The findings of this study were presented to access whether these government portals aligned with the expectations set by Society 5.0.

2.3 Requirements of Society 5.0

To address the limitations, challenges, and applications of next-generation wireless communication systems, the current generation of wireless communication systems must be enhanced to meet the needs of future applications of IoT-based "smart cities." Smart cities, automation, and IoT applications are the real beneficiaries of 5G. The device must be very high in capacity (exceeding 100 Gbps); have low latency, high coverage, and high reliability; and be low in energy, long in battery life, and low in cost.

The high degree of connectivity and speed of data exchange among huge numbers of connected devices has increased demand. 5G wireless networks are considered a vital component of 6G-IoT because they support low-latency use cases at the edge, enabling communication service providers (CSPs) and enterprises to connect mobile IoT devices, data centers, and private and public clouds. As long-term evolution (LTE-A) evolves into 5G, IoT platforms categorized as massive machine-type communications and mission-critical applications should become more prevalent [45].

Additionally, the application of innovative thinking and cognitive skills increase the competitiveness of industries. Through the growth of edge computing (EC), Society 5.0 has minimized latency, reduced network bandwidth, improved privacy and security of data, and facilitated transactions that were previously affected by connectivity issues. It enables societies to exchange information using standard hardware and software resources. By detecting technological issues more quickly, developing better strategies, forecasting future failures, and avoiding big financial losses, the digital twin (DT) helps super-smart cities overcome technological issues. COBOTs are designed to boost productivity while also reinforcing a new human-machine relationship. By removing bottlenecks in communication channels and lowering latency, the Internet of Everything (IoE) gives a chance to minimize operational expenses [46]. To produce and manage massive amounts of data in big data, manufacturers can use data from smart systems and data centers for real-time analysis. As well as enhancing security, blockchains provide authentication and automated service-oriented operations. In the next generation of networks, 6G networks will meet the demands of an intelligent information society while maintaining high energy efficiency, reliability, and low latency.

2.3.1 Anxious Feelings of Inhabitants Regarding the Realization of Society 5.0

In 2016, Japan proposed a super-intelligent society called Society 5.0. This is the Fourth Industrial Revolution, in which advanced technology and connectivity, such as IoT, AI, and big data, will permeate every aspect of society as well as every form of industry. The survey was conducted by the National Institute of Science and

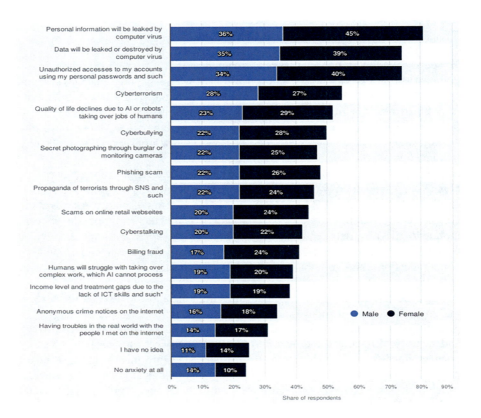

Fig. 2.3 Anxious feeling of inhabitants regarding the realization: a reported survey [47]

Technology Policy (NISTEP) in Japan with 3000 respondents in the age group of 15–69 years as shown in Fig. 2.3. The question was originally posed as follows: What do you feel anxious about regarding the realization of Society 5.0? The research teams used statistics, surveys, and forecasting to collect data based on primary data gathered by their team and secondary data gathered from their collaborators.

In this study, NISTEP investigated the Japanese public's concerns about the implementation of Society 5.0. The majority of respondents worry about their personal information being leaked or hacked by a computer virus, and they are typically concerned about cyberattacks. Approximately 45% of female respondents and 36% of male respondents expressed concern that their personal information would be revealed by a computer virus if the Society 5.0 becomes a reality. The majority of them polled fear cyberbullying and a decline in quality of life due to AI and robots taking over jobs. A lack of ICT skills causes 38% of respondents to be concerned about their income level and treatment gaps. However, approximately 24% of respondents do not have any anxiety about Society 5.0 [47].

2.4 United Nations Sustainable Development Goals (SDGs) for Society 5.0

There were 17 Sustainable Development Goals adopted in 2015 by UN member countries. These goals were aimed at addressing social problems, promoting healthy lives for people, and creating a sustainable world [48, 49]. In 2016, Japan launched the Society 5.0 program, designed to create human-centric and sustainable societies that enhance productivity and quality of life through CPS [50]. In order to achieve the SDGs, social innovation businesses work with customers and partners to address societal challenges [51, 52]. With regard to achieving the SDGs, Keidanren proposes a diverse set of innovations [53]. Following hunter-gatherer societies, agrarian societies, industrial societies, and information societies in history, Society 5.0 is the fifth society. This future society will use IoT, AI, robotics, and other innovative technologies to optimize individual lives and society as a whole. Economic growth combined with solutions to global and local challenges can further enable people of all nationalities, ethnicities, ages, and genders to live comfortably. It adheres to the principles of the UN SDGs. Table 2.2 summarizes the 17 SDGs for Society 5.0 and explains their purpose.

- SDG 1 aims to eliminate severe poverty, provide people with easy access to essential resources and services, and safeguard people from economic and environmental disasters.
- SDG 2 aims to eradicate global poverty and provide safe, nutritious, and sufficient food for all people.
- SDG 3 aims to meet people's basic medical requirements by providing treatment, diagnostics, and medical care in a cost-effective manner.
- SDG 4 will promote accessibility, equality, and lifelong learning for all.
- SDG 5 aims to achieve equality, status, and rights for women and girls in society.
- SDG 6 aims to put new efforts into ensuring that people have access to clean, fresh drinking water.
- SDG 7 encourages energy generation from sources that don't emit CO_2.
- SDG 8 aims to improve economic growth and employment for people as well as ensuring gender equality and providing appropriate compensation.
- SDG 9 promotes people's prosperity by fostering innovations, manufacturing goods to meet human needs, and utilizing industry technologies to construct better, stronger, and more efficient bridges, highways, airports, water, and sewer infrastructure.
- SDG 10 seeks to reduce inequalities within and between countries as well as stop discrimination against older people, individuals with disabilities, women, and racial minorities.
- SDG 11 aims to reduce slums, improve transportation, organize sewage systems, and build more sustainable cities.

Table 2.2 Description of 17 United Nations Sustainable Development Goals [53]

SDGs	Purpose	Goals (by 2030)
SDG 1	No poverty	End poverty in all its forms everywhere
SDG 2	Zero hunger	Ensure to end hunger
SDG 3	Good health and well-being	Ensure healthy lives and promote well-being for all at all ages
SDG 4	Quality education	Ensure that all girls and boys complete free, equitable, and quality primary and secondary education, while substantially increasing the supply of qualified teachers
SDG 5	Gender equality	Achieve gender equality and empower all women and girls
SDG 6	Clean water and sanitation	Ensure access to water and sanitation to all
SDG 7	Affordable and clean energy	Ensure access to affordable, reliable, sustainable, and modern energy
SDG 8	Decent work and economic growth	Promote inclusive and sustainable economic growth, employment, and decent work for all
SDG 9	Industries, innovation, and infrastructure	Build resilient infrastructure and promote sustainable industrialization and faster innovation
SDG 10	Reduce inequalities	Reduce inequalities within and among countries
SDG 11	Sustainable cities and communities	Make cities inclusive, safe, resilient, and sustainable
SDG 12	Responsible consumption and production	Ensure sustainable consumption and production patterns
SDG 13	Climate action	Take urgent action to combat climate change and its impact
SDG 14	Life below water	Conserve and sustainably use the oceans, seas, and marine resources
SDG 15	Life on land	Sustainably manage forests, combat desertification, halt and reserve land degradation, and halt biodiversity loss
SDG 16	Peace, justice, and strong institution	Promote just, peaceful, and inclusive societies
SDG 17	Partnerships for the goals	Revitalize the global partnership for sustainable development

- SDG 12 aims to promote the efficient and cyclical use of food, water, household goods, technological equipment, energy, and all fossil fuels. In the future, the environmental effects of today's products will become more problematic.
- SDG 13 seeks to limit fossil fuel use, create carbon-free cities, and prepare for and respond to natural disasters.
- SDG 14 aims to use seas, oceans, and resources effectively, mitigate for damage, build a healthy structure, and use sustainable resources.
- SDG 15 focuses on maintaining ecosystems and species for future generations.
- SDG 16 aims to create a more just and peaceful community because thousands of civilians die each year in gun battles between nations and rebel groups.
- SDG 17 aims to facilitate successful international cooperation and communication to spur global growth.

2.5 Open Research Challenges and Research Directions 51

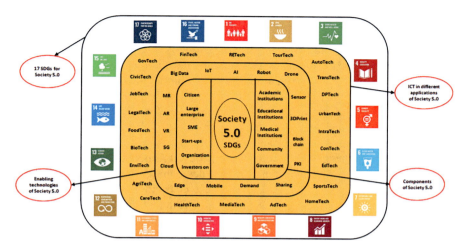

Fig. 2.4 The connection of Society 5.0 with UN SDGs with its components, enabling technologies, and different applications [53]

Society 5.0 with its components and enabling technologies in various technical aspects along with Sustainable Development Goals are well depicted in Fig. 2.4.

2.5 Open Research Challenges and Research Directions

The development of Society 5.0 requires addressing a number of open issues and technical challenges. Several factors are required to build Society 5.0, including energy efficiency, sustainability, expenditures, advanced communication links, defense, and surveillance. Economic, environmental, community, and policy factors influence these factors. Society 5.0 is heavily dependent on financial expenditures, such as design and service costs [54–56]. In contrast to one-time construction costs, operating costs are necessary for a city to maintain its operations. A minimum development cost is necessary for Society 5.0 to become a reality. As a result of low-cost operation, society will be able to function more effectively in the long run. Reduced energy consumption and concerns about environmental waste are the major drivers of the adoption of sustainable smart city solutions. Moreover, governments are implementing stricter regulations to restrict emissions due to concerns regarding global warming and ozone depletion. Consequently, smart grids, intelligent infrastructure automation, and controlling systems ensure a reduction in power consumption and carbon emissions, minimizing losses and optimizing sustainable smart city operations. For sustainable operations, energy efficiency, and cost reduction, carbon emissions and urban garbage need to be significantly reduced in order to achieve sustainability, efficiency, and cost reductions. A sustainable smart city must deal with the growth of pollution and ensure long-term sustainability at the expense of improved operations. Smart city exist because of the efficient use of various smart gadgets that analyze and store massive volume of data.

The IoT, which will play a major part in the emergence of smart cities, requires security and privacy to maintain the sustainability of the ubiquitous city. The deployment of IoT in smart cities continues to present a number of research challenges in the areas of data security and privacy. This is especially true with respect to IoT devices, infrastructure, networks, and protocols. Therefore, creating a sustainable smart city requires considering information security and infrastructure. At the planning and design stage of a sustainable smart city, the safety of residents is a paramount research challenge. Therefore, smart cities are allocating more funds and resources to security, while tech companies are developing solutions that are built to prevent hacking and cybercrime [57–59]. These encryption techniques are being explored by many developers to be used in blockchain-based applications. In Fig. 2.5, open research challenges are presented along with possible solutions.

The Society 5.0 will change the world by giving solutions from various areas. However, there are various potential research challenges, but, in this chapter, we have mainly emphasized the following areas like (1) healthcare, (2) mobility, (3) infrastructure, (4) Fintech.

2.5.1 Healthcare

Healthcare and social security costs are rising, and the elderly are in need of care. By linking and sharing medical data that is currently spread across multiple hospitals, effective medical care will be provided based on data.

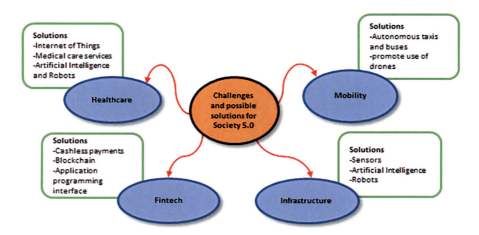

Fig. 2.5 Potential research challenges with possible solutions for Society 5.0

2.5.2 Mobility

Due to a shortage of public transport services, people in unpopulated areas find it difficult to purchase goods and access healthcare. The use of automated vehicles will make it much easier for them to travel, and delivery drones will allow them to get everything they require.

2.5.3 Infrastructure

In addition to ICT, robotics, and sensors for inspections, new technologies can be used to detect sites that require maintenance early. The number of unexpected accidents will be reduced, and construction work time will be reduced, while safety and productivity will be improved.

2.5.4 Fintech

There are limits to the use of information technology in businesses, and cashless payments and convenient financial services are not widely available. The process of sending money to other countries is inconvenient since it takes time and money. With blockchain technology, worldwide commercial transactions will be faster and more secure while promoting cashless payments and saving time.

In the current era of 5G communication technology, several major changes have been observed in the communication and computing environments of the world. Communication and data transfers are smoother and faster due to low-latency connectivity. Additionally, this technology allows quick access to video analytics and AI, which makes society safer. The aim is to enable city and society managers to make informed decisions and provide intelligent public services. Despite the potential that 5G/6G has to transform the world, the smart cities' market is yet to fully realize its potential. A lack of awareness among people is a major factor affecting the development of smart cities, as it causes people to be unaware of smart city projects and reduces government motivation to implement them. Thousands of customers can be securely and reliably served by infrastructure monitoring and management services, irrespective of their location. Regardless of the environment, the service provides high-precision information about the entire infrastructure in real time. Deployment and integration service providers in the industry develop customized products based on user needs, facilitating the integration of smart infrastructure solutions with existing infrastructure by reducing deployment time.

2.6 Summary

Society 5.0 is a super-intelligent society in which emerging advanced technologies are integrated with industry and social life to solve numerous societal concerns and make life more pleasant and sustainable. It envisions a society where everyone enjoys life to the fullest. The purpose of economic growth and technological development is not for the welfare of a select few, but to create prosperity for everyone. In this chapter, we have discussed the potential concepts and technologies behind Society 5.0. The primary purpose is to explore new modalities for efficient usage of the resources and infrastructures of Society 5.0 and fulfill the attributes of the SDGs. In conclusion, we have outlined several open research challenges Society 5.0 faces with effective design. In the context of future projects aimed at improving society the ongoing technological developments worldwide can be regarded as promising initiative. In terms of emerging technologies, 5G-IoT stands out as a highly promising technology with significant potential. The integration of IoT within the framework 5G networks can provide crucial functionalities for Society 5.0, such as efficient resource distribution among residents, identification oversupplies, and real-time information gathering. Considering that societal concerns extend globally, Society 5.0 aims to address the needs and challenges of multiple countries. Also, 5G-IoT enables various sectors of society, including connected health and remote education, while offering enhanced tools for safety and security, thereby enabling the legal system to address issues more effectively.

References

1. https://population.un.org/wup/Publications/Files/WUP2018Highlights.pdf. Accessed on 15 Jan 2023
2. M.M.P. Rana, Urbanization and sustainability: Challenges and strategies for sustainable urban development. Bangladesh. Environ. Dev. Sustain. **13**(1), 237–256 (2011)
3. https://www.marketsandmarkets.com/MarketReports/smart-citiesmarket542.html?Gclid=Cj0KCQiA_JWOBhDRARIsANymNOYaY7mqfAtRT2x3t1D4V2KWNe3vESTj0FRp0CK7IJuBVqwsW78pQaAtIzEALw_wcB. Accessed on 12 Mar 2023
4. G. Culot, G. Nassimbeni, G. Orzes, M. Sartor, Behind the definition of Industry 4.0: Analysis and open questions. Int. J. Prod. Econ. **226**, 1–15 (2020)
5. J. Cotta, M. Breque, L. De Nul, A. Petridis, *Industry 5.0: Towards a Sustainable, Human-Centric and Resilient European Industry* (European Commission Research and Innovation (R&I) Series Policy Brief, 2021)
6. Y. Harayama, Society 5.0: Aiming for a new human-centered society. Hitachi Rev. **66**(6), 556–557 (2017)
7. F. Cugurullo, Exposing smart cities and eco-cities: Frankenstein urbanism and the sustainability challenges of the experimental city. Environ. Plann. A: Econ. Space **50**(1), 73–92 (2018)
8. F. Montori, L. Bedogni, L. Bononi, A collaborative internet of things architecture for smart cities and environmental monitoring. IEEE Internet Things J. **5**(2), 592–605 (2018)
9. W. Li, H. Song, F. Zeng, Policy-based secure and trustworthy sensing for internet of things in smart cities. IEEE Internet Things J. **5**(2), 716–723 (2018)

References

10. B.N. Silva, M. Khan, K. Han, Towards sustainable smart cities: A review of trends, architectures, components, and open challenges in smart cities. Sustain. Cities Soc. **38**, 697–713 (2018)
11. P.M. Santos, C. Queiros, S. Sargento, A. Aguiar, J. Barros, J.G.P. Rodrigues, S.B. Cruz, T. Lourenco, P.M. d'Orey, Y. Luis, C. Rocha, S. Sousa, S. Crisostomo, PortoLivingLab: An IoT-based sensing platform for smart cities. IEEE Internet Things J. **5**(2), 523–532 (2018)
12. C. Harrison, B. Eckman, R. Hamilton, P. Hartswick, J. Kalagnanam, J. Paraszczak, P. Williams, Foundations for smarter cities. IBM J. Res. Dev. **54**(4), 1–16 (2010)
13. A. Deshpande, C. Guestrin, S.R. Madden, J. M. Hellerstein, W. Hong, Model driven data acquisition in sensor networks, in *Proceedings of the 13th International Conference on Very Large Data Bases*, (vol. 30, no. 1, 2004), pp. 588–599
14. https://flex.com/insights/live-smarter-blog/why-5g-will-make-smart-cities-reality. Accessed on 14 Dec 2022
15. D. Bandyopadhyay, J. Sen, Internet of things: Applications and challenges in technology and standardization. Wirel. Pers. Commun. **58**, 49–69 (2011)
16. L. Sanchez, L. Munoz, J.A. Galache, P. Sotres, J.R. Santana, V. Gutierrez, et al., Smart santander: IoT experimentation over a smart city testbed. Comput. Netw. **61**, 217–238 (2014)
17. B.N. Silva, M. Khan, K. Han, Internet of things: A comprehensive review of enabling technologies, architecture, and challenges. IETE Tech. Rev. **35**(2), 205–220 (2018)
18. J. Lee, Performance evaluation of IEEE 802.15.4 for low-rate wireless personal area networks. IEEE Trans. Consum. Electron. **52**(3), 742–749 (2006)
19. Y. Shiroishi, K. Uchiyama, N. Suzuki, Better actions for society 5.0: Using AI for evidence-based policy making that keeps humans in the loop. Computer **52**(11), 73–78 (2019)
20. X. Ge, S. Tu, G. Mao, C.X. Wang, T. Han, 5G ultra-dense cellular networks. IEEE Wirel. Commun. **23**(1), 72–79 (2016)
21. R. Sanchez-Iborra, M.D. Cano, State of the art in LP-WAN solutions for industrial IoT services. Sensors **16**(5), 708/1–708/14 (2016)
22. A. Sołtysik-Piorunkiewicz, I. Zdonek, How society 5.0 and industry 4.0 ideas shape the open data performance expectancy. Sustainability **13**(2), 917/1–917/24 (2021)
23. P. Mishra, P. Thakur, G. Singh, Enabling technologies for IoT based smart city, in *Proceedings of the IEEE, Sixth International Conference on Image Information Processing (ICIIP-2021)*, (India, 26–28 November 2021), pp. 1–6
24. A.V. Anttiroiko, P. Valkama, S.J. Bailey, Smart cities in the new service economy: Building platforms for smart services. AI & Soc. **29**(3), 323–334 (2014)
25. S.P. Mohanty, U. Choppali, E. Kougianos, Everything you wanted to know about smart cities: The internet of things is the backbone. IEEE Consum. Electron. Mag. **5**(3), 60–70 (2016)
26. J. An, F. Le Gall, J. Kim, J. Yun, J. Hwang, M. Bauer, M. Zhao, J. Song, Toward global IoT-enabled smart cities interworking using adaptive semantic adapter. IEEE Internet Things J. **6**(3), 5753–5765 (2019)
27. Y. Lu, Industry 4.0: A survey on technologies, applications and open research issues. J. Ind. Inf. Integr. **6**, 1–10 (2017)
28. A.V. Mavrodieva, R. Shaw, Disaster and climate change issues in Japan's society 5.0—A discussion. Sustainable **12**(5), 1893/1–1893/17 (2020)
29. CeBIT: Japan's Vision of Society 5.0, Euronews. http://www.euronews.com/tag/cebit-2017. Accessed on 12 Jan 2023
30. C. Holroyd, Technological innovation and building a 'super smart' society: Japan's vision of society 5.0. J. Asian Public Policy **15**(1), 18–31 (2020)
31. G. Rathee, A. Sharma, R. Kumar, R. Iqbal, A secure communicating things network framework for industrial IoT using Blockchain technology. Ad Hoc Netw. **94**, 101933/1–101933/15 (2019)
32. R. Poovendran et al., Special issue on cyber-physical systems. Proc. IEEE **100**(1), 6–12 (2012)
33. S.K. Singh, Y.-S. Jeong, J.H. Park, A deep learning-based IoT-oriented infrastructure for secure smart city. Sustain. Cities Soc. **60**, 1–11 (2020)

34. A.G. Pereira, T.M. Lima, F. Charrua-Santos, Industry 4.0 and society 5.0: Opportunities and threats. Int. J. Rec. Technol. Eng. **8**(5), 3305–3308 (2020)
35. A. Deguchi, C. Hirai, H. Matsuoka, T. Nakano, K. Oshima, M. Tai, S. Tani, *What Is Society 5.0? In Society 5.0: A People-Centric Super-Smart Society* (Springer, Singapore, 2020), pp. 1–177
36. A. Iqbal, S. Olariu, A survey of enabling technologies for smart communities. Smart Cities **4**(5), 54–77 (2021)
37. K. Fukuda, Science, technology and innovation ecosystem transformation toward society 5.0. Int. J. Prod. Econ. **220**, 107460/1–107460/14 (2020)
38. D. Serpanos, The cyber-physical systems revolution. Computer **51**(3), 70–73 (2018)
39. T. Higashihara, *A Search for Unicorns and the Building of Society 5.0* (World Economic Forum, Davos, 2018)
40. E.G. Carayannis, J. Draper, B. Bhaneja, Towards fusion energy in the industry 5.0 and society 5.0 context: Call for a global commission for urgent action on fusion energy. J. Knowl. Econ. **12**(4), 1–14 (2021)
41. V. Roblek, M. Mesko, M.P. Bach, O. Thorpe, P. Sprajc, The interaction between internet, sustainable development, and emergence of society 5.0. Data **5**(3), 80/1–80/27 (2020)
42. A. Nikiforova, Smarter open government data for society 5.0: Analysis of 51 OGD portals, (2021). Available online https://zenodo.org/record/5142245. Accessed on 18 Mar 2023
43. K.J.B. Federation, Toward realization of the new economy and society—Reform of the economy and society by the deepening of "Society 5.0". Policy Propos. Ind. Technol. **19**, 1–25 (2016)
44. A. Nikiforova, Smarter open government data for society 5.0: Are your open data smart enough? Sensors **21**(15), 5204/1–5204/28 (2021)
45. https://www.marketsandmarkets.com/Market-Reports/5g-iot-market-14027845.html?gclid=Cj0KCQiA_JWOBhDRARIsANymNOZ6-EnIIRTTV_VE59CRCwz3o_hEaHx-q4fd3aRyoZNlNf1fpb0y0NwsaAjXGEALw_wcB. Accessed on 8 Mar 2023
46. S. Mitchell, N. Villa, M. Stewart-Weeks, A. Lange, *The Internet of Everything for Cities: Connecting People, Process, Data, and Things to Improve the 'Livability' of Cities and Communities* (Cisco, 2013)
47. Reasons why people feel anxious about the realization of Society 5.0 in Japan as of March 2019, by gender [Graph]. Retrieved March 10, 2023, from Statista website https://www.statista.com/statistics/1058259/japan-negative-attitudes-society-5-by-gender/
48. Sustainable Development Goals: 17 Goals to Transform Our World. United Nations. http://www.un.org/sustainabledevelopment. Accessed on 28 Nov 2023
49. A. Opoku, Biodiversity and the built environment: Implications for the Sustainable Development Goals (SDGs). Resour. Conserv. Recycl. **141**, 1–7 (2019)
50. Y. Shiroishi, K. Uchiyama, N. Suzuki, Society 5.0: For human security and well-being. Computer **51**(7), 91–96 (2018)
51. M. Sonmez, N. Suzuki, Towards a society where everyone can enjoy the benefits of new digital technologies: How the fourth industrial revolution is driving the realization of Society 5.0. Hitachi Rev. **68**(2), 5–12 (2019)
52. S. Lamichhane, G. Egilmez, R. Gedik, M.K.S. Bhutta, B. Erenay, Benchmarking OECD countries' sustainable development performance: A goal-specific principal component analysis approach. J. Clean. Prod. **287**, 1–15 (2021)
53. Society 5.0 for SDGs, Keidanren (Japan Business Federation), (8 November 2017). https://www.keidanren.or.jp/en/policy/csr/2017reference2.pdf. Accessed on 25 Jan 2023
54. M.A. Ahad, S. Paiva, G. Tripathi, N. Feroz, Enabling technologies and sustainable smart cities. Sustain. Cities Soc. **61**, 102301/1–102301/12 (2020)
55. H.H. Khan, M.N. Malik, R. Zafar, F.A. Goni, A.G. Chofreh, J.J. Klemes, Y. Alotaibi, Challenges for sustainable smart city development: A conceptual framework. Sustain. Dev. **28**(5), 1507–1518 (2020)

References

56. H. Zahmatkesh, F. Al-Turjman, Fog computing for sustainable smart cities in the IoT era: Caching techniques and enabling technologies—An overview. Sustain. Cities Soc. **59**, 102139/1–102139/15 (2020)
57. H. Kim, H. Choi, H. Kang, J. An, S. Yeom, T. Hong, A systematic review of the smart energy conservation system: From smart homes to sustainable smart cities. Renew. Sust. Energ. Rev. **140**, 110755/1–110755/17 (2021)
58. A. Ahmad, G. Jeon, Y. Chuck Wah, Challenges and emerging technologies for sustainable smart cities. Indoor Built Environ. **30**(5), 581–584 (2021)
59. F. Almalki, S.H. Alsamhi, R. Sahal, J. Hassan, A. Hawbani, N.S. Rajput, A. Saif, J. Morgan, J. Breslin, Green IoT for eco-friendly and sustainable smart cities: Future directions and opportunities. Mob. Netw. Appl. **21**(19), 1–25 (2021)

Chapter 3
Enabling Technologies for Sustainable Smart City

3.1 Introduction

Technology has been used by humanity throughout history to automate processes and make decisions and to ensure a convenient and efficient lifestyle. As the world moves toward Industry 4.0, many cutting-edge technologies are being incorporated into a more efficient and sustainable lifestyle. This is accomplished by implementing the concept of smart cities. The notion of smart cities can now be realized through a number of technological interventions. As a smart city develops, existing and emerging technologies are integrated to support the development of a connected network of devices and entities. It plays a crucial role in enabling various aspects of a smart city, from efficient resources management to improved public services. With this context, Fig. 3.1 presents the key enabling technologies that have applications in a smart city, highlighting their importance and potential impact. Providing seamless connectivity between all the participating entities of a smart city is pivotal to any smart city project. A seamless, self-reliant, and always-on network is achieved through the use of cloud/edge computing paradigms, information and communication technology (ICT), and software-defined networking (SDN) [1, 2]. An automated smart city includes self-sustaining devices embedded with sensors. With the help of cyber-physical systems (CPS), physical devices, services, and management are integrated. The smart city system must provide security and privacy for users and devices.

For providing a security and privacy solution for the devices and participating entities within the smart city system, novel security mechanisms are required. This is because traditional technology and security protocols fail to complement the complex and dynamic nature of the network and devices in the smart city. Consequently, urban areas will be more challenging to live in as the lack of technology in essential sectors such as healthcare, education, the environment, and transportation exacerbates existing problems. To ensure the sustainability of these

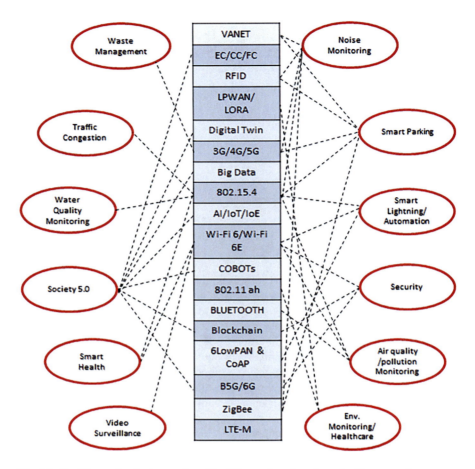

Fig. 3.1 Enabling technologies employed in different domains of sustainable smart city

services, effective methods for data management in urban areas must be developed intelligently. An intelligent city is one that adopts and uses mobile computing systems for these essential services. In order to accomplish this, practical data management networks are developed among all the components and layers of the city [3]. This provide improvements in various aspects, including traffic control, sustainable resources management, quality of life, and infrastructure [4]. As cities become smarter, they increasingly use Internet of Things (IoT), big data, and cloud computing to manage data.

The IoT is a vast network of digital sensors, smart devices, and smart home appliances. The quality of life for citizens will improve as they become more aware of IoT technology. Due to the growing population within cities, adequate services and environmental needs cannot easily be met; therefore, IoT technologies have emerged as a means of creating a smart city that works [5]. It is expected that the rapid development of IoT technologies will motivate scientists and researchers to

3.1 Introduction

create new IoT applications and services [6], and these new smart services will likely meet the needs of citizens worldwide. The exchange and collection of data within IoT services will also be taken into account to raise awareness of smart city concepts throughout the world. The network must therefore be embedded with actuating, networking, computing, and sensing functions [7]. A smart city is developed and analyzed by monitoring, gathering, archiving, and sharing open sensor data from IoT devices. There is a wide variety of literature on smart city key themes, including environmental monitoring for smart cities [8]; people's quality of life in smart cities that focuses mainly on four different phenomena at the city scale, namely, weather, environment, public transportation, and people flow; and semantic web environments for aggregation and quality analysis in smart cities [9].

To improve urban smartness, a connection between physical, social, business, and technology infrastructure is discussed in [10]. A modern city is defined in [11] as one that uses ICT to improve its quality of life and quality of urban services. This technology enhances the everyday operations of city life as well as the performance of individual citizens [12]. A smart city concept or theme (such as smart people, smart transportation, and more) is needed as a result of the evolution of ICT technologies. A city's components and layers need to be connected deeply using data management technologies, i.e., IoT, big data, cloud computing, and more. As shown in Fig. 3.2, various communication, computing, sensing and other wireless technologies domains are all involved in sustainable smart cities applications and play an important role in a range of applications by adopting enabling technologies such as edge computing (EC), digital twin (DT), Internet of Energy (IoE), big data analytics, COBOTs, 6G, blockchain, etc.

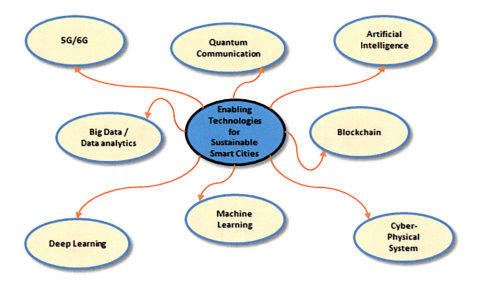

Fig. 3.2 Key enabling technologies for sustainable smart city

In preliminary data analysis, Fig. 3.3a illustrates the documents published per year with the number of publications. The most publications are made in the year 2021 with 40, and as publications are growing each year, this is an excellent field for further research. Articles are the document type with a large number of publications with a percentage of 57.7%. Figure 3.3b shows the document per year by source with subject area. *IEEE Access* has more publications i.e. more than six. Computer Science subject domain has most of the publications with a percentage of 33.8%, and Engineering has the second most publications with a percentage of 23.3%. Figure 3.3c shows the congregation of keywords that are occurring together in this chapter. The interrelations between keywords that are co-appearing in the research are linked together. Internet of Things, smart city, 5G, big data, and blockchain are the most co-appearing keywords.

Hence, smart cities bring together ICT with existing city infrastructure to coordinate and manage it using digital technology. Smart cities are slowly but surely becoming a reality as many countries around the world are adopting this idea and developing their own models. Sensors and actuators embedded in smart devices sense the environment to make effective decisions in a smart city. In order to accomplish this, several technologies must be integrated, including artificial intelligence (AI), protocols, the IoT, and wireless sensor networks (WSN).

A modern utopian world that balances traditional systems with highly advanced technological interventions has been made possible through the cutting-edge technological advancements of today. Smart cities are one such promising project that has been adopted globally with an aim to enhance the lives of its inhabitants [13]. The idea is to use modern-day technologies to convert every entity of a conventional city into an autonomous object performing its operation automatically without substantial external assistance. The ability to access these anywhere in the world is made possible with the help of smart devices. Smart cities are realized through the collection of substantial data from a variety of sources within an urban environment [14].

Smart devices and their recent advancements have made it highly desirable to connect everyday objects via existing networks. As a result of the evolution of conventional networks, a huge number of devices have been connected to the IoT. The IoT's concept has been further strengthened by technological advancements in ubiquitous computing (UC), WSN, and machine-to-machine (M2M) communication [15, 16]. In IoT, UC is enabled through uniquely identifiable smart devices with minimal or no human interaction [17]. To support contextual decision-making, connected smart devices share their own information and access authorized information from other devices [18]. The use of IoT technology has pioneered an array of striking applications, including intelligent smart homes, smart cities, intelligent warehouses, smart health, and more [19–21]. Smart cities are applications of the IoT, so they inherit the same operating mechanisms. In this chapter, we discuss extensively the role that enabling technologies play in smart cities. We have presented most promising key enabling technologies extensively and their requirements to be developed for sustainable smart cities. These include 5G/6G, quantum communication, IoT, IoE, AI, machine learning (ML), deep learning (DL), big data analytics, blockchain, CPS and wireless sensor networks.

3.1 Introduction

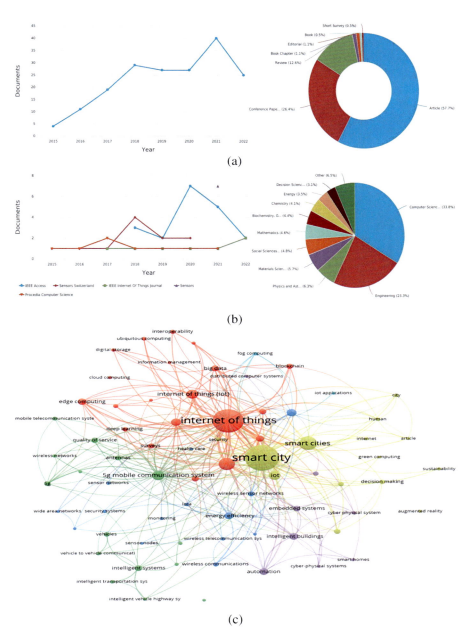

Fig. 3.3 Preliminary data analysis. (**a**) Documents published per year with the number of publications. (**b**) Document per year by source with subject area. (**c**) The congregation of co-appearing keywords from the Scopus database

The rest of the chapter is organized as follows. The recent advances in communication technologies are presented in Sect. 3.2. The Internet of Things/Internet of Everything is discussed in Sect. 3.3. Emerging technologies like artificial intelligence, machine learning, deep learning, and big data/data analytics are explained in Sects. 3.4 and 3.5, respectively. Blockchain, cyber-physical system, and wireless sensor networks were presented in Sects. 3.6, 3.7, and 3.8. The potential research challenges and research directions are explored in Sect. 3.9. Finally, Sect. 3.10 concludes the chapter and recommends the future scope.

3.2 Recent Advances in Communication Technologies

3.2.1 5G/6G

As smart cities become more integrated, 5G networks offer increased flexibility and allow information to be shared more quickly. With enhanced connectivity, more data is collected and analyzed, promoting digital change in smart cities. It is possible to create smart urban environments through 5G through advanced infrastructure, sustainable development, accessibility, and equality [22]. Although 5G is not officially launched yet, researchers have turned their attention to the 6G communication system. Future mobile networks should support various unknown IoT services, that is, the model and functionality of the network must adapt to the ever-changing features and requirements of the services.

In other words, the future network should be intelligent, capable of learning the characteristics of the service autonomously and being aware of their changes. Additionally, it should be designed in a way that the architecture and functionalities can be easily updated as changes occur. There is a threefold increase in data rate in 6G, ten times the latency in the quickest 5G network. In addition, there is a hundred times higher level of reliability. 6G will be capable of connecting everything; integrating multiple technologies; supporting holographic, tactile, aerial, and marine communications; as well as supporting the Internet of Everything, Internet of Nano-Things, and Internet of Bodies. Applications like drones and robots connected to these IoT devices will provide a variety of services, such as smart traffic, environment monitoring and control, telemedicine, digital sensing, and high-definition (HD) and full HD video transmission. Satellite networks will also be used in 6G for global coverage. Satellite networks is divided into three types: telecommunication satellite networks, navigation satellite networks, and earth imaging satellite networks. With 6G Internet, we will be able to connect to the internet at a very fast speed using a combination of radio technology and fiber-optic technology. By delivering internet through line of sight on a 6G network, speed is not impacted by distance.

3.2.2 Quantum Communication

Over the past few years, quantum computing has gained a lot of attention because of its supremacy. It shows that a quantum computer solves a problem in a time-efficient manner, which ordinary computers cannot achieve, using a programmable quantum device. This has been illustrated by Google, Honeywell, and other organizations where quantum computing research is in progress. The development of a smart city can be facilitated by the application of quantum computing technologies. It enables real-time decision-making and data analysis, making it highly valuable for smart city initiative. The use of quantum computing in smart cities is anticipated to be a highly effective commercial application of this technology. It is particularly well-suited for scenarios that involve large-scale data processing, prioritizing efficiency over absolute precision. These types of challenges are expected to be prevalent in future smart cities.

Utilizing the power of quantum computing, each smart car would be equipped with AI algorithms that not only solve local transportation problems, but also leverage the immense computational capabilities to optimize traffic flows, ensuring seamless and efficient mobility throughout the city. To make real-time decisions on whether to stop or accelerate at intersections, smart cars continuously scan their environment. However, such decisions on a large scale may not be optimal. Therefore, the city could benefit from utilizing a quantum computer to optimize traffic flows throughout the cities, suggesting different routes for different cars to reduce travel time. Another potential application of quantum computing is the optimization of electrical grids, where a small error can be tolerated but requires significant computational power.

3.3 Internet of Things (IoT)/Internet of Everything (IoE)

During the 1990s, internet-based technologies began to be adopted, leading to the concept of the "Internet of Things" [23, 24]. By connecting physical and virtual things through ICT, [25] defines the IoT as a technology that enables the provision of advanced services on a global scale. IoT devices enable users to interact with even more complex concepts of automated systems [26, 27]. IoT allows multiple devices to communicate simultaneously through smart computing, embedded devices, and sensor networks. Basically, the IoT consists of a network of intelligent devices that communicate with one another. Since the advent of the IoT, cloud technology has been effectively integrated with sensors used in a wide range of applications.

IoT can be categorized into two main areas: technology and applications [28]. With the help of several sensory methods, users are capable of interacting with real-time information [29], while the technology domain characterizes IoT network design based on the underlying network, namely, service quality [30]. Additionally,

IPv6-enabled architecture, flexible IoT-categorized design frameworks [31], and energy management IoT systems are mainly used in automation systems. Also, different research papers have been published in the application domain [32–34]. There are various applications of IoT in modern cities, such as cybervilles, digital cities, information cities, smart cities, and wired cities [35]. There are also a variety of applications that have been developed for IoT, including IP cameras [36], smart wheelchairs [37], building information management, early warning systems based on cloud services [38], and the Web of Things (WoT) [39, 40]. The "Internet of Everything" is a concept that extends the IoT by connecting not just devices and sensors but also people, processes, and data in a comprehensive network. In the context of a smart city, the IoE plays a significant role in integrating various components to enable a more efficient and interconnected urban environment. It involves the seamless connectivity of devices, infrastructure, services, and citizens, enabling real-time data collection, analysis, and decision-making. Leveraging IoE technologies, smart cities enhance several aspects of urban life. For instance, in transportation, connected vehicles and infrastructure optimize traffic flow, reduce congestion, and improve safety. In public services, smart grids efficiently manage energy consumption, while connected waste management systems optimize collection routes and reduce environmental impact. Additionally, in healthcare, IoE devices and systems enable remote patient monitoring, enhance emergency response, and improve overall healthcare services. In a smart city context refers to the interconnected and integration of various elements within the urban landscape, including devices, infrastructure, services, and people, with the goal of improving efficiency, sustainability, and quality of life. Within smart city applications, a variety of approaches have been envisioned due to recent studies that have addressed the potential of IoT/IoE in the design of sustainable smart cities.

3.4 Artificial Intelligence (AI)/Machine Learning (ML) /Deep Learning (DL)

An excessive amount of data produced by a smart city is useless unless it is analyzed in order to extract useful information from it. In a smart city setup, AI simplifies the processing and analysis of machine-to-machine communication information. By using ML and DL technologies, it is possible to create predictive and preventative decisions. Consequently, AI, ML, and DL technologies have made significant contributions to the development of smart cities. These advanced technologies are revolutionizing various aspects of urban life. Intelligent traffic management systems powered by AI and ML algorithms have transformed transportation efficiency, reducing congestion and optimizing traffic flow. Energy management has become more sustainable and efficient with ML algorithms analyzing data from smart grids and integrating renewable energy sources. Predictive maintenance using ML and DL algorithms has improved infrastructure management by detecting potential failures and enabling proactive maintenance. AI-powered surveillance systems enhance public safety and security by detecting anomalies and identifying potential threats in real-time. Waste management has become more efficient and environmentally friendly with ML algorithms optimizing waste collection routes and schedules. Smart governance and service delivery benefit from AI-powered chatbots and virtual assistants, improving citizen engagement and streamlining administrative processes. Lastly, ML algorithms enable environmental monitoring, helping cities monitor air quality, detect environmental hazards, and predict climate patterns. In addition, it is possible to gain a holistic understanding of intra- and intersystem settings [41–43]. Overall, AI, ML, and DL technologies play crucial roles in creating sustainable, efficient, and livable smart cities.

3.5 Big Data/Data Analytics

Developing sustainable smart cities requires the use of big data because it allows cities to draw valuable information from a large amount of data from various sources [44, 45]. Data management effectively enhances services for the public, minimizes management costs, and maximizes resource efficiency [46, 47]. Smart cities necessitate the integration of such a huge amount of data from a wide variety of sources. Integration of data is one of the most challenging aspects of smart cities. It is not just a technical challenge to achieve data integrity when dealing with information from various sources, but it is also crucial to understand socio-technical aspects like privacy and power. As a result of the convergence of ICTs enabling real-time insight into the physical world and big data analytics, smart cities are empowered to make intelligent decisions about strategies to maximize efficiency and effectiveness [48]. The development and management of smart cities depends on intelligence, connectivity, and information sharing [49].

Smart cities use big data to provide better services to the public, track climatic conditions, increase transportation efficiency, lower pollution levels, and offer better city security services. Urban situations are able to be analyzed more precisely and in real time by means of big data analysis. In addition, big data allows cities to extract useful information from a huge amount of data obtained from a wide range of sources, providing useful information for decision-making, promoting sustainability, and offering citizens new services [50–55].

However, big data can be used by sensors and devices to generate valuable information. As a result of the availability of large computational and storage facilities for processing these data, the application of big data in smart cities has many advantages and limitations. Its reliance on cloud computing and IoT services, however, is one of its greatest advantages. The most common implementation of big data in smart cities requires intelligent networks that connect components and residents' equipment, such as smartphones and cars. Therefore, quality of service is very significant for real-time big data applications in smart cities [56–61].

3.6 Blockchain

IoT devices are seamlessly interconnected to one another through blockchain technology, making it possible to manage data distributed across a network. Each device on the network securely access and exchange data with other devices through a decentralized record of transactions stored on the blockchain. This makes it possible for IoT devices to communicate with one another without any human intervention, providing a secure, efficient, and cost-effective way to manage data. With blockchain integrated into smart city environments, information is exchanged safely and robustly. There is tremendous potential for this technology to be a viable concept for applications in smart cities. This makes it an ideal solution for applications in smart cities, as it ensures data security, accuracy, and reliability [62, 63].

3.7 Cyber-Physical Systems (CPS)

A cyber-physical system integrates sensing, computation, control, and networking into physical objects and infrastructure. It consists of entities, devices, networks, processing, management, computations, and related physical processes. These systems use sensors to monitor physical processes, actuators to control physical processes, and software to integrate, control, and analyze the data collected. They are designed to interact with the physical world and is used to monitor and control complex systems, such as robotics, manufacturing, transportation, healthcare, and energy systems. Essentially, it includes all computing, processing, networking, and storage functions [64]. Additionally, CPS addresses the challenges of resilience and security in smart cities. Robust cyber security measures are implemented to protect critical infrastructure from cyber threats and ensure the privacy and integrity of data. It fosters citizen engagement by providing platforms for real-time information sharing, citizen feedback, and participation. This involvement allows residents to actively contribute to urban planning, public service delivery, and community development, making smart cities more inclusive and responsive to the needs of their inhabitants. With CPS as the backbone, smart cities can harness the power of technology to create more efficient, sustainable, and livable urban environments for their residents.

3.8 Wireless Sensor Networks (WSN)

In order for smart cities to be successful, WSNs are crucial for efficient data sensing, interpretation, and communication. In a smart city, they are also used for planning traffic, controlling temperature, and detecting location [65–67]. WSNs are necessary for smart cities because they allow for the efficient collection and transmission of data. This data is used to monitor and control the environment, optimize traffic flow, and detect locations that can help improve safety and efficiency. It plays a vital role in smart cities by enabling real-time data collection, monitoring, and analysis. They facilitate efficient resource management, enhance decision-making, and contribute to the overall sustainability and livability of the urban environment.

In addition to healthcare and government services, WSNs produce a lot of useful data and can be applied in many different ways [68]. By integrating a WSN with RFIDs technology, it becomes feasible to obtain valuable data about individuals, objects, and their movements. This combination enables the WSN to collect information from the surrounding environment, while RFIDs provide identification and tracking capabilities. Further, an important component in this process is the analog-to-digital converter (ADC), which utilizes sensors to convert analog signals into digital signals [69]. An ADC converts analog data into digital data according to the wireless sensor node framework, which includes various types of analog sensors. A memory and microcontroller are responsible for executing specific operations on data according to the data requirements. An interface for radio transmission completes the transmission of data. It is necessary to equip all of this equipment with a power supply. In its fully developed state, a WSN consists of compact, energy-efficient, and cost-effective sensor nodes that can be deployed in diverse environments and function reliably over extended periods of time.

3.9 Open Research Challenges and Research Directions

Implementing an effective and sustainable smart city involves several inherent and external challenges and issues. It plays a major role in building trust and achieving acceptance in a smart city, as security is the primary concern of all stakeholders. The widespread acceptance of smart city projects has been seriously hindered by several cases of data theft and cybercrime in recent years. Smart city projects are hindered by high implementation costs caused by high adoption costs of evolving technologies. Implementation costs for upgrading existing infrastructure are high. A smart city relies on the communication of valuable information between various IoT devices. In order for a smart city to survive, it is imperative that these devices are interoperable. In order to effectively manage and process the massive amounts of data generated by billions of locally connected devices, intelligent techniques must be employed along with deeper and wider insights. A transition from legacy systems to intelligent systems is further complicated by technical challenges. Backward compatibility, scalability, heterogeneity of data and devices, multiple data standards, and interoperability pose inherent challenges.

3.10 Summary

Several developing nations still envision the possibility of achieving smart cities, but with modern technological advancements, these nations have a platform in place to move in this direction. Innovative technologies are being developed to meet the dynamic and diverse needs of smart cities. These technologies include autonomous vehicles, renewable energy systems, IoT, advanced analytics, and AI. With these technologies, emerging nations are able to create more efficient, resilient, and sustainable cities. This is done by improving access to public services, reducing pollution and traffic congestion, and improving urban planning. Security is one of the major pillars of a smart city. Technological advances in AI and IoT allow for better monitoring and control of public spaces to ensure safety and security. It also allows for the collection of data to improve urban planning and decision-making, as well as the development of smart services that can be used to address problems in a smart city system.

Cyber-physical systems play a crucial role in the sustainability of a smart city. A wide range of cyber-physical systems frameworks and architectures are currently being developed around the globe by the researchers. For a smart city, companies and organizations are developing cost-effective, inclusive, and environmentally friendly solutions based on the Sustainable Development Goals. Therefore, smart cities should make use of various enabling technologies to create transparent, secure, adaptable, and manageable cities. In future research, nanotechnology and quantum computing must play a significant role in achieving sustainability.

References

1. M.A. Ahad, R. Biswas, Request-based, secured and energy-efficient (RBSEE) architecture for handling IoT big data. J. Inf. Sci. **45**(2), 227–238 (2019)
2. M.A. Ahad, G. Tripathi, S. Zafar, F. Doja, IoT data management—Security aspects of information linkage in IoT systems, in *Principles of Internet of Things (IoT) Ecosystem: Insight Paradigm*, (Springer, Cham, 2020), pp. 439–464
3. D. Gavalas, P. Nicopolitidis, A. Kameas, C. Goumopoulos, P. Bellavista, L. Lambrinos, B. Guo, Smart cities: Recent trends, methodologies, and applications. Wirel. Commun. Mob. Comput. **2017**, 1–2 (2017)
4. E. Ismagilova, L. Hughes, Y.K. Dwivedi, K.R. Raman, Smart cities: Advances in research-An information systems perspective. Int. J. Inf. Manag. **47**, 88–100 (2019)
5. J. Jin, J. Gubbi, S. Marusic, M. Palaniswami, An information framework for creating a smart city through Internet of Things. IEEE Internet Things J. **1**(2), 112–121 (2014)
6. M. Mohammadi, A. Al-Fuqaha, M. Guizani, J.-S. Oh, Semi-supervised deep reinforcement learning in support of IoT and smart city services. IEEE Internet Things J. **5**(2), 624–635 (2018)
7. W. Li, H. Song, F. Zeng, Policy-based secure and trustworthy sensing for Internet of Things in smart cities. IEEE Internet Things J. **5**(2), 716–723 (2018)
8. F. Montori, L. Bedogni, L. Bononi, A collaborative Internet of Things architecture for smart cities and environmental monitoring. IEEE Internet Things J. **5**(2), 592–605 (2018)
9. J. An, F. Le Gall, J. Kim, J. Yun, J. Hwang, M. Bauer, M. Zhao, J. Song, Toward global IoT-enabled smart cities interworking using adaptive semantic adapter. IEEE Internet Things J. **6**(3), 5753–5765 (2019)
10. C. Harrison, B. Eckman, R. Hamilton, P. Hartswick, J. Kalagnanam, J. Paraszczak, P. Williams, Foundations for smarter cities. IBM J. Res. Dev. **54**(4), 1–16 (2010)
11. S. Kondepudi, Smart sustainable cities analysis of definitions, in *ITUT Focus Group Smart Sustainable Cities*, Technical report, (United Nations, Washington, DC, 2014)
12. B.N. Silva, M. Khan, K. Han, Towards sustainable smart cities: A review of trends, architectures, components, and open challenges in smart cities. Sustain. Cities Soc. **38**, 697–713 (2018)

13. T. Nam, T.A. Pardo, Conceptualizing smart city with dimensions of technology, people, and institutions, in *Proceedings of the 12th Annual International Digital Government Research Conference: Digital Government Innovation in Challenging Times*, (June 2011), pp. 282–291
14. M. Batty, Big data, smart cities and city planning. Dialogues Hum. Geogr. **3**(3), 274–279 (2013)
15. B.N. Silva, M. Khan, K. Han, Internet of things: A comprehensive review of enabling technologies, architecture, and challenges. IETE Tech. Rev. **35**(2), 205–220 (2017)
16. M. Khan, B.N. Silva, C. Jung, K. Han, A context-Aware smart home control system based on ZigBee sensor network. KSII Trans. Internet Inf. Syst. **11**, 1057–1069 (2017)
17. J. Gubbi, R. Buyya, S. Marusic, M. Palaniswami, Internet of Things (IoT): A vision, architectural elements, and future directions. Futur. Gener. Comput. Syst. **29**, 1645–1660 (2013)
18. S. Vongsingthong, S. Smanchat, Internet of things: A review of applications and technologies. Suranaree J. Sci. Technol. **21**(4), 359–374 (2014)
19. S.R. Islam, D. Kwak, M.H. Kabir, M. Hossain, K.-S. Kwak, The Internet of Things for health care: A comprehensive survey. IEEE Access **3**, 678–708 (2015)
20. A. Zanella, N. Bui, A. Castellani, L. Vangelista, M. Zorzi, Internet of things for smart cities. IEEE Internet Things J. **1**(1), 22–32 (2014)
21. M. Khan, B.N. Silva, K. Han, Internet of things based energy aware smart home control system. IEEE Access **4**, 7556–7566 (2016)
22. B. Detwiller, *5G Will Bring Smart Cities to Life in Unexpected Ways* (2020). Available at https://www.techrepublic.com/article/5g-will-bring-smart-cities-to-life-inunexpected-ways/. Accessed on 10 Mar 2023
23. A.H. Alavi, P. Jiao, W.G. Buttlar, N. Lajnef, Internet of Things-enabled smart cities: State-of-the-art and future trends. Measurement **129**, 589–606 (2018)
24. K. Ashton, That 'Internet of Things' thing. RFID J. **22**(7), 97–144 (2009)
25. H. Sundmaeker, P. Guillemin, P. Friess, S. Woelffle, European commission, and directorate-general for the information society and media, vision and challenges for realising the Internet of Things. Cluster Eur. Res. Proj. Internet Things Eur. Commission **3**(3), 34–36 (2010)
26. C. Perera, C.H. Liu, S. Jayawardena, The emerging Internet of Things marketplace from an industrial perspective: A survey. IEEE Trans. Emerg. Top. Comput. **3**(4), 585–598 (2015)
27. C. Perera, A. Zaslavsky, P. Christen, D. Georgakopoulos, Context aware computing for the Internet of Things: A survey. IEEE Commun. Surv. Tutorials **16**(1), 414–454 (2014)
28. A. Al-Fuqaha, M. Guizani, M. Mohammadi, M. Aledhari, M. Ayyash, Internet of Things: A survey on enabling technologies, protocols, and applications. IEEE Commun. Surv. Tutorials **17**(4), 2347–2376 (2015)
29. R. Jalali, K. El-khatib, C. McGregor, Smart city architecture for community level services through the Internet of Things, in *Proceedings of the 18th International Conference on Intelligence in Next Generation Networks*, (2015), pp. 108–113
30. J. Jin, J. Gubbi, T. Luo, M. Palaniswami, Network architecture and QoS issues in the Internet of Things for a smart city, in *Proceedings of the International Symposium on Communications and Information Technologies (ISCIT)*, (2012), pp. 956–961
31. A. Ghasempour, Optimum number of aggregators based on power consumption, cost, and network lifetime in advanced metering infrastructure architecture for smart grid Internet of Things, in *Proceedings of the 13th IEEE Annual Consumer Communications & Networking Conference (CCNC)*, (2016), pp. 295–296
32. R. Petrolo, V. Loscrí, N. Mitton, Towards a smart city based on cloud of things, in *Proceedings of the ACM International Workshop on Wireless and Mobile Technologies for Smart Cities WiMobCity*, (2014), pp. 61–66
33. M. Cesana, A.E.C. Redondi, IoT Communication technologies for smart cities, in *Designing, Developing, Facilitating Smart Cities*, ed. by V. Angelakis, E. Tragos, H.C. Pohls, A. Kapovits, A. Bassi, (Springer, Cham, 2017), pp. 139–162
34. H. Salehi, R. Burgueno, Emerging artificial intelligence methods in structural engineering. Eng. Struct. **171**, 170–189 (2018)
35. S.P. Mohanty, U. Choppali, E. Kougianos, Everything you wanted to know about smart cities: The Internet of Things is the backbone. IEEE Consum. Electron. Mag. **5**(3), 60–70 (2016)

36. C. Gonzalez Garcia, D. Meana-Llorian, B.C.P. G-Bustelo, J.M.C. Lovelle, N. Garcia-Fernandez, Midgar: Detection of people through computer vision in the Internet of Things scenarios to improve the security in smart cities, smart towns, and smart homes. Futur. Gener. Comput. Syst. **76**, 301–313 (2017)
37. Z. Rashid, J. Melia-Segui, R. Pous, E. Peig, Using augmented reality and Internet of Things to improve accessibility of people with motor disabilities in the context of smart cities. Futur. Gener. Comput. Syst. **76**, 248–261 (2017)
38. J. Wang, C. He, Y. Liu, G. Tian, I. Peng, J. Xing, X. Ruan, H. Xie, F.L. Wang, Efficient alarm behavior analytics for telecom networks. Inf. Sci. **402**, 1–14 (2017)
39. A. Gyrard, M. Serrano, Connected smart cities: Interoperability with SEG 3.0 for the Internet of Things, in *Proceedings of the 30th International Conference on Advanced Information Networking and Applications Workshops (WAINA)*, (2016), pp. 796–802
40. A. Gyrard, P. Patel, S.K. Datta, M. Intizar, Semantic Web meets Internet of Things (IoT) and Web of Things (WoT), in *Proceedings 15th International Semantic Web Conference (ISWC)*, vol. 10, (2016), pp. 1–11
41. Z. Allam, Z.A. Dhunny, On big data, artificial intelligence and smart cities. Cities **89**, 80–91 (2019)
42. N. Komninos, C. Kakderi, A. Panori, P. Tsarchopoulos, Smart city planning from an evolutionary perspective. J. Urban Technol. **26**(2), 3–20 (2019)
43. K.E. Skouby, P. Lynggaard, Smart home and smart city solutions enabled by 5G, IoT, AAI and CoT services, in *Proceedings of International Conference on Contemporary Computing and Informatics (IC3I)*, (2014, November), pp. 874–878
44. R. Matheus, M. Janssen, D. Maheshwari, Data science empowering the public: Data-driven dashboards for transparent and accountable decision-making in smart cities. Gov. Inf. Q. **37**(3), 101284/1–101284/9 (2020)
45. I.A.T. Hashem, V. Chang, N.B. Anuar, K. Adewole, I. Yaqoob, A. Gani, E. Ahmed, H. Chiroma, The role of big data in smart city. Int. J. Inf. Manag. **36**(5), 748–758 (2016)
46. E. Al Nuaimi, H. Al Neyadi, N. Mohamed, J. Al-Jaroodi, Applications of big data to smart cities. J. Internet Serv. Appl. **6**(1), 1–15 (2015)
47. M. Batty, Big data and the city. Built Environ. **42**(3), 321–337 (2016)
48. H. Chourabi, T. Nam, S. Walker, J. R. Gil-Garcia, S. Mellouli, K. Nahon, et al., Understanding smart cities: An integrative framework, in *Paper presented at the System Science (HICSS), 2012 45th Hawaii international conference on*, (TBD Maui, HI, USA, 2012)
49. R. Khatoun, S. Zeadally, Smart cities: concepts, architectures, research opportunities. Commun. ACM **59**(8), 46–57 (2016)
50. B. Allen, L.E. Tamindael, S.H. Bickerton, W. Cho, Does citizen coproduction lead to better urban services in smart cities projects? An empirical study on e-participation in a mobile big data platform. Gov. Inf. Q. **37**(1), 101412/1–10 (2020)
51. K.C. Desouza, B. Jacob, Big data in the public sector: Lessons for practitioners and scholars. Adm. Soc. **49**(7), 1043–1064 (2017)
52. R. Kitchin, The real-time city? Big data and smart urbanism. GeoJournal **79**(1), 1–14 (2014)
53. D. Zhang, S.L. Pan, J. Yu, W. Liu, Orchestrating big data analytics capability for sustainability: A study of air pollution management in China. Inf. Manag., 103231 (2019). https://doi.org/10.1016/j.im.2019.103231
54. R. Diaz-Diaz, L. Munoz, D. Perez-Gonzalez, Business model analysis of public services operating in the smart city ecosystem: The case of Smart Santander. Futur. Gener. Comput. Syst. **76**, 198–214 (2017)
55. M. Wazid, A.K. Das, R. Hussain, G. Succi, J.J.P.C. Rodrigues, Authentication in cloud-driven IoT-based big data environment: Survey and outlook. J. Syst. Archit. **97**, 185–196 (2019)
56. H. Cai, B. Xu, L. Jiang, A.V. Vasilakos, IoT-based big data storage systems in cloud computing: Perspectives and challenges. IEEE Internet Things J. **4**(1), 75–87 (2017)
57. M. Marjani, F. Nasaruddin, A. Gani, A. Karim, I.A.T. Hashem, A. Siddiqa, I. Yaqoob, Big IoT data analytics: Architecture, opportunities, and open research challenges. IEEE Access **5**, 5247–5261 (2017)

References

58. A. Jindal, A. Dua, N. Kumar, A.K. Das, A.V. Vasilakos, J.J.P.C. Rodrigues, Providing Healthcare-as-a-Service using fuzzy rule based big data analytics in cloud computing. IEEE J. Biomed. Health Inform. **22**(5), 1605–1618 (2018)
59. M. Talebkhah, A. Sali, M. Marjani, M. Gordan, S.J. Hashim, F.Z. Rokhani, IoT and big data applications in smart cities: Recent advances, challenges, and critical issues. IEEE Access **9**, 55465–55484 (2021)
60. A. Gani, A. Siddiqa, S. Shamshirband, F. Hanum, A survey on indexing techniques for big data: Taxonomy and performance evaluation. Knowl. Inf. Syst. **46**(2), 241–284 (2016)
61. N. Khan, I. Yaqoob, I.A.T. Hashem, Z. Inayat, W.K.M. Ali, M. Alam, M. Shiraz, A. Gani, Big data: Survey, technologies, opportunities, and challenges. Sci. World J. **2014**, 1–18 (2014)
62. M. Swan, *Blockchain: Blueprint for a new economy* (O'Reilly Media, 2015)
63. S. Underwood, Blockchain beyond bitcoin. Commun. ACM **59**(11), 15–17 (2016)
64. E.A. Lee, Cyber physical systems: Design challenges, in *Proceedings of 11th IEEE International Symposium on Object and Component-Oriented Real-Time Distributed Computing (ISORC)*, (May 2008), pp. 363–369
65. M.S. Jamil, M.A. Jamil, A. Mazhar, A. Ikram, A. Ahmed, U. Munawar, Smart environment monitoring system by employing wireless sensor networks on vehicles for pollution free smart cities. Proc. Eng. **107**, 480–484 (2015)
66. J. Wu, K. Ota, M. Dong, C. Li, A hierarchical security framework for defending against sophisticated attacks on wireless sensor networks in smart cities. IEEE Access **4**, 416–424 (2016)
67. J. Yick, B. Mukherjee, D. Ghosal, Wireless sensor network survey. Comput. Netw. **52**(12), 2292–2330 (2008)
68. A. Alamri, W.S. Ansari, M.M. Hassan, M.S. Hossain, A. Alelaiwi, M.A. Hossain, A survey on sensor-cloud: Architecture, applications, and approaches. Int. J. Distrib. Sens. Netw. **9**, 917923/1–18 (2013)
69. G. Hancke, B. Silva Jr., G. Hancke, The role of advanced sensing in smart cities. Sensors **13**, 393–425 (2012)

Chapter 4
Internet of Things for Sustainable Smart City

4.1 Introduction

Everyday objects of all kinds are enticipated to become integral components of a complex and interconnected system, forming a sophisticated network of devices and physical entities. This encompasses various elements includes citizens, traffic, highways, transportation systems, buildings, sewage systems, electrical networks, automobiles, machinery, merchandise, plants, soil, and water. The Internet of Things (IoT) is transforming the way data is collected, shared, and analyzed by establishing an interconnected network. By connecting physical objects to the internet, it allows for the exchange of data between the two, enabling devices to respond to each other and their environment. This data exchange enables users to remotely monitor and control physical objects, leading to automation, improved efficiency, and the development of new services and applications. Moreover, it enable devices to learn from each other, fostering intelligent and adaptive behavior. Through the IoT, humans, devices, tools, and cities become interconnected, allowing cities to collect data from sensors, cameras, and other devices to gain insights into city operation, human behavior and environment interactions. This data then be used to optimize the city's infrastructure and services, making it more efficient and cost-effective. Additionally, the IoT helps improve public safety, provide better access to city services, and enable more efficient environmental management. The purpose of the IoT is to perform intelligent tasks by sharing information and communication. These tasks include identifying, tracking, finding, searching, monitoring, controlling, evaluating, managing, operating, and planning things [1].

With the growing popularity of IoT and related big data applications become more popular, smart sustainable cities are becoming a reality. A range of technological and scientific advancements in ubiquitous computing and big data analytics have contributed to this development, including multi-sensor data fusion, hybrid reasoning and modelling, machine learning, data mining, cloud computing, wireless communication networks, etc. Smart sustainable cities are emerging as a holistic urban development strategy and a transformative phenomenon driven by information intelligence integrated into urban systems. It opens up with entirely new opportunities for smart cities and sustainable cities to explicitly incorporate environmental sustainability. In terms of operations, functions, designs, and services, smart sustainable cities represent transformational processes driven by information intelligence embedded into urban systems. By enhancing and integrating urban systems and facilitating coordination and collaboration among diverse urban domains, such intelligence enabled and driven by IoT and its underlying big data analytics can advance environmental urban sustainability. Data science, computer science, and complexity science, upon which the IoT is based, are increasingly recognized as promising answers to the challenge of environmental urban sustainability. Smart sustainable cities must address the complex mechanisms and patterns involved in the interactions between environmental and physical systems [2]. The applications of IoT depicted in Fig. 4.1 is used to build sustainable smart cities and create a more sustainable and livable environment for citizens.

In preliminary data analysis, Fig. 4.2a illustrates the number of documents published per year with the number of publications. The most publications are made in the year 2021 with 207, and as publications are growing each year, this is an excellent field for further research. Articles are the document type with a large number of publications with a percentage of 42.2%. Figure 4.2b shows the document per year by source with subject area. *Sustainable Cities and Society* has more publications more than 15. Computer Science subject domain has most of the publications with a percentage of 28.3%, and Engineering has the second most publications with a percentage of 22.4%. Figure 4.2c shows the congregation of keywords that are occurring together in a paper. The interrelations between keywords that are co-appearing in the research are linked together. Internet of Things, sustainable smart

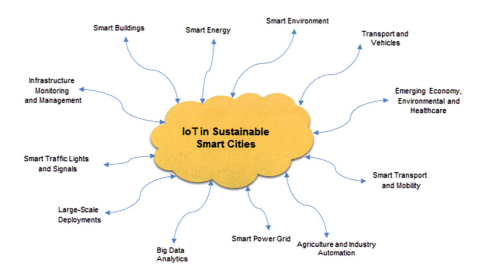

Fig. 4.1 Concept of IoT in sustainable smart cities

4.1 Introduction

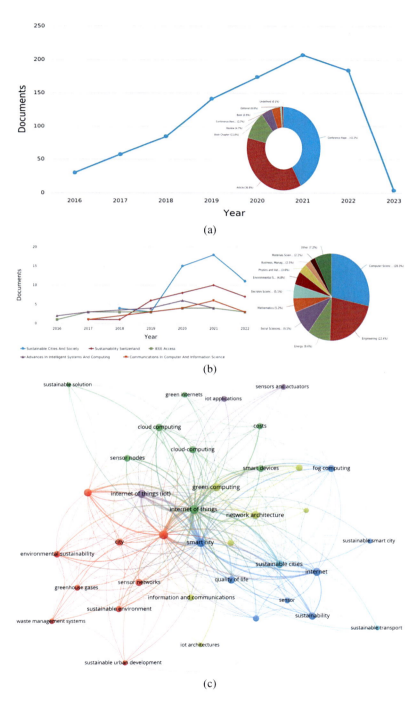

Fig. 4.2 Preliminary data analysis. (**a**) Documents published per year with the number of publications, (**b**) document per year by source with subject area, and (**c**) the congregation of co-appearing keywords from the Scopus database

city, IoT architecture, IoT applications, fog computing, and cloud computing are the most co-appearing keywords. The rest of the chapter is organized as follows. IoT Architecture for Sustainable Smart Cities are presented in Sect. 4.2. Computing Technologies in IoT Environment is discussed in Sect. 4.3. IoT Applications in Sustainable Smart Cities is discussed in Sect. 4.4. The potential research challenges and research directions are explored in Sect. 4.5. Finally, Sect. 4.6 concludes the chapter and recommends the future scope.

4.2 IoT Architecture for Sustainable Smart Cities

The IoT architecture is designed to enable cities to become more sustainable by utilizing technology such as sensors, actuators, and controllers to connect and monitor different parts of the city. This allows data to be collected and analyzed to help cities reduce energy consumption, optimize waste management, and improve the quality of life for citizens. This architecture is designed to provide a framework for the integration of IoT devices and systems into a sustainable smart city. It includes data collection, data storage, data analytics, and decision-making processes that is used to monitor and manage the city's resources more efficiently. By leveraging the power of the IoT, cities become more sustainable, reduce their environmental impact, and improve the quality of life for their citizens. To realize the vision of a sustainable smart city, it is essential to address several technological, business, legal, and regulatory challenges. These challenges encompasses ensuring scalability, reliability, security, and liability in a wide range of application scenarios, particularly in light of the projected 68% urban population by 2050 [3]. Properly addressing these issues is crucial to create a robust and sustainable infrastructure that meets the evolving needs of urban areas. Businesses and governments are increasingly investing in 5G-IoT technologies, with spending predicted to reach USD 158 billion in 2022 in order to cope with such massive urban population growth [4]. As part of the sustainable smart city services, IoT sensors and networks will be used to capture data for big data analytics, with edge computing ensuring ultra-fast and reliable data transfer and 5G networks connecting sensors and smart assets [5].

By 2026, the global 5G-IoT market will reach USD 40.2 billion, a 73% CAGR (compound annual growth rate) over USD 2.6 billion in 2021. Due to its high processing speeds, massive capacity, and super low latency, 5G is predicted to be the strongest enabler of IoT expansion. A 5G-IoT network support a wide range of static and mobile communication devices, including those with different speeds, bandwidth, and quality-of-service requirements. Due to its large bandwidth, massive scalability, high reliability, low latency, and unprecedented speed, the 5G network enables faster adoption of applications like autonomous vehicles, artificial intelligence (AI), and virtual reality (VR) / augmented reality (AR) for consumer applications that require high data rates and rapid responses [6]. As shown in Fig. 4.3, we have upended the architecture, providing an in-depth analysis of several existing architecture models.

4.2 IoT Architecture for Sustainable Smart Cities

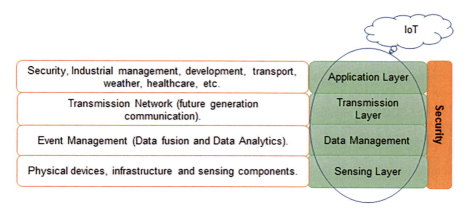

Fig. 4.3 ICT architecture with IoT for sustainable smart city

There are four main components of the presented architecture:

- Application layer
- Transmission layer
- Data management layer
- Sensing layer

Further, the privacy and security of the information are also very important concerns for the sustainable smart city. Therefore, the securities devices have to be integrated into each of the layers. The information collection of the physical devices is the main task of sensing devices, which is located near the bottom of the structure. The communication devices assist to communicate information to the next layer.

4.2.1 Application Layer

The application layer is the topmost level of the sustainable smart city architecture that works between city residents and information storage management. The performance has a very strong impact from the user's point of view, which directly deals with the citizen himself. Most importantly, the services provided include the smart transport, the development of the community, and more. Thus, facilitating the sharing of data via different applications is seemed to be favorable for the development of a society.

4.2.2 Transmission Layer

The transmission layer plays a vital role in supporting the development of the sustainable smart city. It essentially act as a convergence point for various communication networks [7], incorporating a combination of wired, wireless, and satellite technologies. In terms of coverage, it is classified into access and network transmission. ZigBee, near-field communication (NFC), bluetooth, machine-to-machine (M2M) communication, and Z-Wave are examples of access network technologies that provide small range coverage. The same technique to provide wider coverage in transmission network technologies are 3G/4G LTE/5G, LoRA, and low-power wide-area network (LPWAN). For the identification of any object or person, Radio-frequency identification (RFID) technology makes use of a part of the electromagnetic spectrum. However, it does a good job of identifying all of the things seen from afar, similar to system of bar codes.

NFC is a network technology that allows two devices to communicate and identify each other over a 10 cm distance. Bluetooth wireless technology is a well-known technology for access to the network, which greatly reduces the power consumption of the communication on the part of short-wave radio signals. ZigBee is self-organized, is reliable, and offers communication at a very low power within a range of 10 m with an expendable battery [8]. LTE has been the leading-edge wireless technology, a 4G network, which has proven its advantage over 3G, Wi-Fi, and WiMAX, in terms of increased bandwidth, faster data transfer speeds, and low latency. However, the fifth generation is a prominent technology that supports massive MIMO to transfer bulk data in gigabit [9]. Cellular service providers use licenses and bandwidth to offer mobile services, and Wi-Fi works on unlicensed frequency bands offering high bandwidth. LPWAN is a promising and novel transmission networking model for smart cities intent to improve the energy efficiency of industrial networking [10] for wide-area coverage at low power. Light fidelity (Li-Fi) is another trend in data transmission techniques that uses ambient light-emitting diodes (LEDs) to transmit data. Based on the availability of spectrum, as well as the potential of high-speed communication, Li-Fi is considered as the future of the wireless communication network.

4.2.3 Data Management Layer

The data management layer serves as the "brain" of a sustainable smart city, positioned between the application layer and the acquisition layer. It is responsible for analyzing, organizing, storing, manipulating, and making decisions based on the collected data. The primary role of the data layer is to ensure the integrity and viability of information, with a focus on data maintenance and cleaning.

4.2.4 Sensing Layer

The lowest layer of the IoT is the sensing layer, which consists of wireless sensor network, smart devices, and data capturing devices used to collect data from devices and sensors. It offers various methods [11] to improve the performance of data collection and storage in various frameworks depending upon various parameters like pressure, temperature, light, humidity, and more. The RFID sensors, cameras, GPS, actuators, Bluetooth, and ZigBee are different sensing equipment used by experts [12]. Mostly based on literature, the intelligence of cities is growing along with the

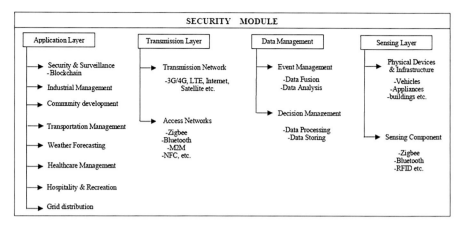

Fig. 4.4 The security module of IoT architecture for sustainable smart cities

occupancy of sensor networks [13]. Further, the sensing layer collects information from different mediums and infrastructure, resulting in an increasing number of connected devices in the sustainable smart city network. The security module of IoT architecture for sustainable smart cities with applications is shown in Fig. 4.4.

4.3 Computing Technologies in IoT Environment

This section discusses cloud computing, edge computing, fog computing, and quantum computing in the context of IoT. All of these technologies have the potential to increase the efficiency and effectiveness of IoT networks. Cloud computing is used to store and process data from IoT devices, while edge computing help reduce latency by processing data directly on the edge of the network. Fog computing bridge the gap between cloud and edge computing, and quantum computing enhance the speed and accuracy of AI algorithms. The cloud offers a centralized way for devices to manage their data, but there are challenges to consider. A cloud-based solution may not always be suitable, and certain cases require rapid response time. In such situations, there are two main approaches to distributing computational capacity: edge computing and fog computing. It is important to be aware of hidden costs associated with complex solutions like these, and evaluating specific use cases can help determine the viability of edge computing for businesses.

The IoT is often portrayed as a network of many devices distributed around the world, connected to the cloud in an almost magical way. There are many challenges associated with moving huge amounts of data from the machines we connect to the cloud. However, in many cases, we don't get what we want despite solving the challenges. The information collected should be used in a timely manner to enhance its value. Therefore, it is not always the right solution to take advantage of large, centralized data by connecting things to the cloud. In this context, cloud, edge, fog, and quantum computing are discussed.

4.3.1 Cloud Computing

A centralized data lake in the cloud is used to store cold data consolidated in distributed computing. They are used to analyze them at another level, typically for more strategic purposes. Cloud computing is a good solution for processing big data. With a centralized data lake in the cloud, organizations store large volumes of data in a single location. This makes the data easier to access and manage, and it also reduces the cost of data storage. Additionally, the cloud provides access to powerful computing resources at a lower cost, which makes it ideal for processing big data. It is a paradigm that offers users access to a common platform of computational resources. The use of cloud computing, for instance, offer both organizations and individuals various benefits in terms of cost savings and expenditure reductions [14].

It provides the opportunity to share resources and calculations from a certain location through a network-based environment. The template connects to a network of servers, storage systems, software, and programmable grids, which is easily set up via communication with a provider [15]. The users of cloud computing are unaware of the exact location of their sensitive data because different security challenges and risks exist in physical data centers. Because of the fast propagation of threats in virtualized systems, traditional security solutions such as firewalls and host-based antivirus software are not suitable security platforms [14].

A further benefit of cloud computing is the immediate availability of hardware resources as well as lower upfront costs and cost-effective maintenance [16]. As an enabler, cloud computing also improve operational activities and agility [17]. Cloud computing can be viewed from two different perspectives, namely, the technical perspective and the client perspective [17]. From a technical point of view, security, virtualization, architecture, and pricing models all play an important role [18]. In [19] study, switching costs, security, and privacy issues were identified within the cloud infrastructure. Cloud computing and smart connectivity are supported in [20] as well as real-time applications for smart cities. Additionally, a framework was presented for integrating real and virtual devices with highly scattered, assorted, and reorganized data. In terms of operating platforms, software, and infrastructure, this technology has greatly contributed to cost-effective, faster solutions. Physical infrastructure does not need to be set up by users; rather, it can be purchased as a service from cloud service providers [21, 22].

4.3.2 Edge Computing

Edge computing is a distributed computing paradigm that moves data processing and applications closer to the edge of a network, near the data source. This reduces latency, increases privacy, and reduces the cost of computation. It is particularly useful for modifying real-time machine behaviors based on data. There are production lines in many factories that handle different elements until they are assembled and prepared for shipment. It is essential to perform quality control before packaging on these production lines. It is not always possible to identify which of the systems or steps of the line are failing merely by observing the problem. Therefore, each step must be closely monitored. This data is processed on the local network, simplifying the manager's work in supervising and improving the plant. A factory cannot afford to send data to the cloud, apart from the sheer volume of information. Thus, edge computing would be useful for tracking a production line.

Through edge computing, users are able to perform quick, lightweight computations right on the edge of a network rather than having to transfer the entire data to the cloud [23, 24]. The system can then take decisions in near real time as a result of the faster response time. This reduces the amount of data transferred to the cloud, which improves the scalability of the system. Moreover, it allows the system to take decisions locally, without the need to wait for the response from the cloud, thus improving the overall response time. This allows the system to process data faster and make decisions without relying on the cloud, which reduces latency. This means the system respond to events almost immediately, providing better performance and scalability. However, edge computing reduces the amount of data that needs to be sent to the cloud, which help reduce costs associated with cloud computing.

4.3.3 Fog Computing

Fog computing is a model of cloud computing that is closer to the ground, meaning that instead of having data stored and processed on a central server, it is stored and processed on distributed edge devices, such as routers and gateways, which are closer to the user. This allows for faster and more efficient data processing. Fog computing in smart cities helps reduce latency and improve the user experience. It also helps reduce the cost of data storage and processing by removing the need for a central server. Additionally, it improves reliability and reduce the risk of data breaches as the data is stored and processed on multiple edge devices. By utilizing fog computing in smart cities, companies benefit from improved speed, efficiency, and cost savings—all while ensuring data safety and reliability. For example, in the healthcare industry, fog computing is used to store and process patient data securely at the edge of the network, enabling faster analysis and decision-making while ensuring privacy and data safety. For instance, a hospital could use fog computing to store patient records and medical images at the edge of the network, allowing doctors to access patient data quickly and securely without the need to transfer large amounts of data over the network. However, there are also potential risks associated with fog computing. For example, if data is stored at the edge of the network, it may be more vulnerable to attack. Additionally, fog computing can add complexity to the network, which may make it more difficult to manage and troubleshoot.

4.3.4 Quantum Computing

Quantum computing is a type of computing where information is processed with quantum bits instead of classical bits. This makes quantum computers much faster and more powerful than traditional computers. It is an emerging form of computing that uses the principles of quantum mechanics to process and store data. It uses quantum bits (qubits) instead of conventional bits to represent and manipulate data, allowing for much faster and more efficient calculations. This is because qubits exist in multiple states at once, allowing them to store and process a huge amount of information in a very short amount of time. This makes it possible to solve problems that would take traditional computers a very long time to solve. This makes quantum computing incredibly powerful and opens up opportunities for ground-breaking advancements in areas such as artificial intelligence, machine learning, and cryptography. As a result, quantum computing revolutionize the way we process information and allow us to develop innovative solutions to complex problems. Smart cities rely heavily on data and technology, and quantum computing provide the necessary power to process and analyze vast amounts of data quickly and efficiently. Additionally, quantum computing helps with the development of more efficient and improved algorithms for problem-solving, which is applied to optimize urban planning and management. Finally, quantum computing also helps create more secure networks and encryption methods, which are essential for protecting the data of citizens living in smart cities. This is due to the fact that quantum computing has the ability to process immense amounts of data at incredibly fast speeds, which makes it ideal for solving complex problems. Furthermore, the ability to process data securely is essential for protecting the privacy of citizens in smart cities, as well as for ensuring the integrity of the data being collected.

However, quantum computing faces two main challenges: (1) quantum computers are highly prone to interference that leads to errors in quantum algorithms running on them, and (2) most quantum computers cannot function without being super cooled to a temperature slightly above absolute zero, as heat generates errors or noise in qubits.

4.4 IoT Applications in Sustainable Smart Cities

IoT-enabled smart cities allow for more efficient use of resources through the use of sensors, which is used to monitor and control energy, water, and waste management systems. They also provide a better quality of life for citizens by providing real-time data about traffic, air quality, and other environmental conditions. In addition, IoT-enabled smart cities promote sustainability by allowing for more efficient use of resources and improving the quality of life for citizens. Among them are transport, mobility, traffic lights, traffic grids, energy systems, buildings, infrastructure monitoring and management, and planning [25].

4.4 IoT Applications in Sustainable Smart Cities

- **Smart Transport and Mobility**

The use of IoT technologies in transportation and mobility helps to improve efficiency, reduce costs, and optimize traffic flows. IoT enables the tracking of vehicles and the monitoring of roads, bridges, and other infrastructure, allowing for increased safety and better management of transportation systems. IoT-based smart transport and mobility smart city is a system that uses the IoT technology to improve the efficiency of transportation and mobility services in urban areas. This system makes use of sensors, connected devices, and data analytics to provide an integrated view of all transportation- and mobility-related activities. It helps to reduce traffic congestion, optimize energy consumption, and improve road safety.

IoT applications related to environmental sustainability are found in transport and mobility. Intelligent transportation and mobility are key components of smart sustainable cities. The uses of IoT encompass a wide range of applications, including automating the tracking of individuals and their vehicles, monitoring traffic congestion and road conditions, identifying available parking spaces, implementing pricing and tolling mechanisms, and overseeing the distribution of valuable goods [26–29]. In addition, the IoT provides advanced location-based services related to onboard navigation systems that allow effective use of existing transport infrastructure and networks through wirelessly transmitted and displayed multimedia presentations. Additionally, this helps drivers select cost-efficient and time-efficient routes. In addition, GPS and the IoT is used to gather information and make predictions regarding pollution density [30].

- **Smart Traffic Lights and Signals**

IoT-based smart traffic lights and signals allow the traffic flow to be managed more efficiently. They use sensors and cameras to detect traffic flow and adjust the speed of traffic and the timing of traffic signals accordingly. This helps reduce congestion, increase safety, and improve air quality. Transport, commuting, and other forms of mobility are also enhanced by improved traffic patterns. In order to achieve this, a high volume of traffic congestion will be handled by measuring a variety of parameters using a variety of sensors. Providing better traffic services through smart traffic lights and signals requires sensors and computational mechanisms that function properly. A traffic grid should interconnect these lights and signals, data should be collected from all traffic lights at various spatial scales, and intelligent decision systems built on real-time big data should be implemented. As explained in [31], IoT can reduce road congestion and accidents by opening new roads, directing vehicles to alternative roads, collecting and providing parking lot information, and improving transportation infrastructure based on congestion data.

- **Smart Energy**

Smart energy systems enable the integration of renewable energy sources, such as solar and wind, into the city's energy grid. They also enable improved energy efficiency through the monitoring of energy usage patterns and the implementation of demand-responsive energy management strategies. This helps to reduce energy costs and emissions, as well as creating a more reliable and resilient energy system. By leveraging IoT powered by big data, power supply decisions are made based on citizen demand. Energy companies and decision-makers respond to market fluctuations based on real-time data. This involves determining when to increase or decrease production, which contributes to energy efficiency. Using the same technologies enables near-real-time forecasting and prediction based on large datasets collected and stored.

The IoT-enabled mobile devices provide citizens with live energy prices and enable them to adjust their use, accordingly, reducing the burden on energy costs [32]. In order to align with energy resource optimization as strategic objectives, specific pricing plans that are based on supply, demand, production, and consumption models is used [31]. Through the integration of sensing and actuation systems, the IoT also optimizes and controls energy consumption [33]. With an integrated system, all types of appliances and energy-consuming devices are integrated, and real-time data is collected and transferred to utilities, enabling them to balance power generation, distribution, and usage effectively [34].

- **IoT-Based Smart Power Grid**

This technology is based on the use of sensors, monitors, and other devices connected to the internet to track and manage electricity usage. It allows utilities to identify areas of high consumption and manage energy more efficiently by shifting the load to other areas. It also helps to reduce energy costs as it is more efficient than traditional methods.

Using smart grid technology, big data is collected from diverse power sources and processed and analyzed in a real-time manner to improve power system performance by transmitting relevant information [35, 36]. In addition, the IoT offers huge potential for power grid management since it allows systems to gather and act on near-real-time data on energy consumption, generation, and inefficiency from end users. Sensors are placed on consumer access points as well as on production, transmission, and distribution systems as part of a smart metering infrastructure. In terms of monitoring customers' energy consumption, smart meters have been recognized for their efficiency.

With the IoT, new trade mechanisms are built on the basis of supply and demand in the energy market [37]. As a result of smart metering, it is possible to avoid costly and carbon-intensive peaks in the power grid by using new means of coordination as to the overall ensemble of consumers, it enables aggregation of real-time data on energy consumption and the definition of dynamic pricing schemes [37]. The use of advanced technologies in renewable energy enables power grids to improve planning and coordination around the generation of power from renewable plants.

- **IoT-Based Smart Environment**

IoT-based smart environment is a network of connected devices that are able to collect data, analyze it, and take automated actions based on the results. This enables more efficient use of energy and resources, as well as improved safety and security. By utilizing IoT-based smart environment, businesses and individuals achieve

4.4 IoT Applications in Sustainable Smart Cities

greater operational efficiency and security, while ensuring sustainability through the efficient use of resources.

Air quality and noise pollution are reduced through the IoT, enhancing the quality of life for citizens. To achieve this, air quality sensors are mounted on bike wheels and cars and deployed throughout the city. By measuring the pollutants in the air, these sensors adjust traffic flows, manage parks, and even measure emissions from individual buildings. This data informs policy decisions and create incentives to reduce emissions and noise levels. A machine learning and data mining model is proposed in [38], which integrates old data, mobility data, air data, speed data, and traffic data for further processing with artificial neural networks and conditional random fields. Sensors are used for monitoring air quality and atmospheric conditions, assisting in environmental protection through IoT [39]. By detecting pollution in the air and water, ubiquitous sensors help reduce pollution harmful to public health [40].

- **IoT-Based Smart Buildings**

Smart buildings enable the integration of various digital systems and technologies to enable efficient management of building operations such as energy, security, and maintenance. This results in an improved user experience, enhanced security, and cost savings. In residential, industrial, public, and commercial buildings, the IoT uses sensors and actuators to monitor and control mechanical, electrical, and electronic systems. In smart buildings, it controls, monitors, and adjusts the heating, ventilation, and air conditioning (HVAC), lighting, home automation, and power systems, all mechanical and electrical components and devices automatically.

This way, it manages the building environment, monitors system performance, controls demand, controls temperature, reduces heat loss, tracks carbon emissions, manages windows and doors, provides lighting based on occupancy schedules, etc. Three major areas of literature are covered here:

- Building energy management systems with IoT devices to create smart buildings that are energy-efficient and IoT-driven.
- Monitoring occupant behavior and energy consumption in real time.
- A future application of smart devices in the built environment.

Physical and digital objects with embedded sensors and actuators can process information, self-configure, and make independent decisions about their operation [41]. A self-configuring IoT application adapts to the surrounding environment, an intelligent behavior that is triggered independently to deal with emerging conditions [42].

- **Infrastructure Monitoring and Management**

By having a comprehensive IoT-based infrastructure in place, cities can monitor and manage their urban infrastructure more efficiently. This allows them to maximize the efficiency and availability of resources while also reducing costs associated with maintenance and repair. This in turn leads to improved public services and overall better quality of life for citizens. Using IoT, urban infrastructure such as bridges, railway tracks, and tunnels are monitored and controlled [43]. It refers to risk and cost increases in urban infrastructures as well as compromises in safety and quality. In this respect, IoT devices improves incident management, emergency response coordination, and service quality and save money on infrastructure maintenance [44]. By coordinating tasks between different service providers and users of these infrastructures and facilities, the IoT infrastructure allows for scheduling repair and maintenance activities more efficiently [25]. Furthermore, waste and water management, as well as distributed networks, constitute key IoT applications.

- **Typology and Design Concepts for Sustainable Urban Forms**

Smart sustainable cities utilize big data analytics enabled by the IoT to improve their contribution to environmental sustainability goals by using analytics enabled by the IoT. In order to investigate and evaluate sustainable urban form typologies and design concepts in terms of their contributions to environmental sustainability, it improves data collection, analysis, modelling, and simulation [25].

This has affected urban practices in terms of applying environmental sustainability principles and methods in urban planning and development. For investigating sustainable urban forms, big data analytics is a useful alternative to traditional data collection and analysis methods. Big mobility data is used to evaluate environmental performance, and city simulation models are used to assess and optimize sustaining urban forms' contributions to sustainability. In the future, big data analytics will be used in sustainable urban forms.

- **Large-Scale Deployments**

The IoT is being deployed on large scales around the world. This has been made possible by the development of more powerful and efficient processors, low-cost and low-power sensors, and wireless communication technologies. These advancements have enabled the cost-effective deployment of millions of devices in homes, businesses, and cities. Big data processing platforms and computing platforms are developed to optimize operations, management, and coordination of urban planning [45]. A number of countries are utilizing the IoT to enable smart cities, which are increasingly considered national urban development projects that promote the use of the IoT and big data analytics in sustainable urban development. Meanwhile, sustainable cities across the globe are being wired, connected, and networked and generating continuous streams of data collected and processed for different decision-making purposes. Smart cities are able to use data to optimize energy and water usage, reduce waste and pollution, and create more efficient transportation systems. As cities become smarter, they become more sustainable and more resilient in the face of climate change.

Increasingly, large-scale deployments of the IoT have been justified by the benefits they provide for environmental sustainability. Large-scale deployments include weather monitoring, reducing noise, sustainability, smart buildings, tracking controlling infrastructure, traffic control, shipping, smart parking, and e-ticketing.

Smart sustainable cities significantly improve the environmental dimension of sustainability through the use of IoT and big data analytics. The use of this

4.4 IoT Applications in Sustainable Smart Cities

technology and its big data applications is based on integrating and harnessing solutions and approaches through coordination, coupling, and merging urban systems and domains. Smart sustainable cities offer a wide range of benefits and opportunities as a result of exposing big data via an evolvable, dynamic, extensible, scalable, and reliable IoT ecosystem.

- **Economic Development, Environmental Protection, and Healthcare**

IoT is used to improve environmental sustainability and healthcare access in emerging economies. It is used to monitor air quality, water quality, and soil quality in an emerging economy. Moreover, it is utilized to track the spread of infectious diseases, monitor the health of vulnerable populations, and provide remote medical care. The IoT is completely committed to providing benefits and achieving the UN's Sustainable Development Goals in the areas of health, education, and economics. Another important issue is the sustainability of the environment. IoT developers should take environmental factors into consideration when developing IoT devices and systems. IoT devices consume a lot of energy, which is one of the challenges associated with their environmental impact. It is also possible to adapt green technologies in order to create efficient energy-efficient devices for the future. Furthermore, it is beneficial for human health as well as the environment. Several health issues such as diabetes, obesity, and depression are being monitored with highly efficient IoT devices [46].

- **Industrial Automation and Agriculture**

Automation has been used in the industrial sector for decades, but the application of automation technology to the agricultural sector is relatively new. Automation is used to increase efficiency and accuracy in crop planting, harvesting, and other processes, resulting in higher yields and lower costs. Agriculture must be advanced in order to feed such a huge population. For this reason, agriculture and technology need to be combined to improve production efficiency. Greenhouse technology is one of the possible approaches in this direction. This allows the production process to be improved by controlling environmental parameters. Smart devices and sensors enable more efficient climate control and process monitoring inside the chamber due to the advancement of IoT. Factory automatization, warehouse management, quality management, logistics, and supply chain management can be automated, providing innovative solutions.

- **Big Data Analytics**

IoT devices produce large amounts of data that is used to gain valuable insights into customer behavior and trends. It consists of many devices and sensors that communicate with one another. These sensors and devices are rapidly increasing in number as IoT networks expand and grow. A massive amount of data is transferred over the internet by these devices communicating with each other. By using big data analytics, businesses better understand their customers, optimize their products and services, and increase their profits. It is used to collect and analyze data from sensors in IoT-enabled smart cities. This data is then used to create insights that is used to improve the efficiency and performance of public services such as transportation, energy, and waste management.

In smart buildings, IoT big data frameworks are very useful to monitor oxygen concentrations, detect smoke and toxic gases, and determine luminosity [47]. The use of IoT-based cyber-physical systems equipped with information analysis and knowledge acquisition techniques also improves industrial production [48]. An IoT-based traffic management system analyzes real-time traffic information collected by IoT devices and sensors installed in traffic signals [49]. It is crucial for healthcare systems to integrate large amounts of information into one database and process it in real time to take quick, accurate decisions [50]. Using IoT and big data analytics, manufacturing industries also modernizes their traditional approaches [51]. In addition, cloud computing and analytics reduce costs and increase customer satisfaction in the development and conservation of energy [52]. This large amount of information can be effectively handled by deep learning, and it can provide results with high accuracy [53]. Therefore, IoT and big data analytics are therefore very critical for the development of smart cities.

4.5 Open Research Challenges and Research Directions

Various technologies involved in data transfer between embedded devices and the involvement of IoT-based systems in all aspects of human life made it complex and raised several issues and challenges. These challenges include designing secure and resilient architectures, developing effective data management systems, creating efficient and reliable communication protocols, developing context-aware applications, and tackling privacy and trust issues. Additionally, IoT developers face these challenges as we move toward a smarter society with advanced technology. With the expansion of technology, challenges and the need for IoT systems are also increasing. It is therefore necessary for IoT developers to provide solutions to the new issues that arise as the IoT develops.

- **Security and Privacy Issues in IoT**

IoT devices collect and transmit sensitive personal data, and the security of the data is often not taken into account. Without proper security measures, hackers access the data and use it for malicious purposes. Since these devices are often connected to the internet, they are vulnerable to cyberattacks [54]. All of these issues leads to unauthorized access to user data and the potential for personal data to be stolen or manipulated. Poor transport layer encryption also makes it easier for attackers to intercept and modify communications between the device and the user [55]. To develop IoT systems, security and privacy issues are very critical with regard to various aspects [56], including encryption methods. To prevent attacks and threats, IoT architecture must incorporate security mechanisms at every layer. To ensure security and privacy in IoT-based systems, several protocols are developed and efficiently deployed across all communication channels [57, 58].

4.5 Open Research Challenges and Research Directions

It is imperative to note, however, that some IoT applications require different methods to ensure the security of their communication. Additionally, if wireless technologies are used in the IoT system, security risks are increased. The detection of malicious actions and self-healing or recovery should be achieved using certain methods. On the other hand, privacy is another crucial concern that users should consider when using IoT solutions. The communication between trusted parties must therefore be established over a secure network that provides authorization and authentication [59]. The IoT system also has different privacy policies for different objects communicating with each other. Consequently, each IoT object should be able to verify the privacy policies of other IoT objects before transmitting data.

- **Interoperability/Standard Issues**

The concept of interoperability refers to the ability of different IoT devices and systems to exchange information between themselves. It is important to keep in mind that this exchange of information does not rely on the hardware or software that has been deployed. In order to develop an IoT system that is interoperable, we have to address the heterogeneity of the technologies and solutions used in the process. Interoperability is divided into four levels based on technical, semantic, syntactic, and organizational interoperability [60]. In order to improve the interoperability of IoT systems, various functionalities are being provided to ensure you are able to communicate with different objects in a heterogeneous environment by providing various functionalities. The use of different IoT platforms on the basis of their functionalities can also be combined. This approach provides various solutions for IoT users based on the functionality of these platforms [61]. There are several approaches to handling interoperability, which are also called interoperability handling approaches, which have been approved by researchers as a solution to this important issue [62]. For instance, these solutions could be available as adapters/gateways, virtual networks/overlays, service-oriented architectures, and more. Despite the fact that interoperability handling approaches ease some of the pressure on IoT systems, there are still certain challenges remaining with interoperability that could be the subject of future research.

- **Ethics, Law, and Regulatory Rights**

The ethics, law, and regulatory rights of IoT developers are also important issues to consider. For the purpose of maintaining standards and moral values and preventing people from violating them, there are certain rules and regulations that are in place. A distinction between ethics and law is that ethics are standards that people believe and laws are government-imposed restrictions. Nevertheless, both ethics and laws aim to maintain the standard and prevent illegal use. Several real-life problems have been solved as a result of the development of IoT in the past few years. However, IoT has also been a catalyst for critical challenges in terms of ethics and law [63]. There are many challenges associated with the use of data, including security, privacy protection, trust and safety, and data usability. In addition, the majority of IoT users have also been observed to be supportive of government norms and regulations in regard to data protection, privacy, and safety as a result of the lack of trust they have in IoT devices. Thus, it is of the utmost importance that this issue is taken into consideration in order to maintain and improve the trust of people when using IoT devices and systems in the future.

- **Scalability, Availability, and Reliability**

System scalability means the ability to add new services, equipment, and devices without degradation of performance. In IoT, the main challenge is supporting many different devices with various memory, processing, storage, and bandwidth requirements. There is also the matter of the availability of the service that needs to be considered [64, 65]. As part of the layered IoT framework, scalability and availability should both be implemented together. IoT cloud systems offer scalability by adding devices, storage, and processing power as required, allowing the IoT network to scale as needed. The development of an IoT framework that meets global needs is possible with this globally distributed IoT network. There is also the challenge of ensuring that resources are available to authentic objects regardless of where they are located or when they need them. To utilize the resources and services of global IoT platforms, several small IoT networks are timely connected to them.

In order to ensure the availability of some services and resources, different data transmission channels, such as satellite communication, may be used. The availability of resources and services will be uninterrupted only if there is an independent and reliable channel of data transmission.

- **Quality of Service**

Quality of services (QoS) is a crucial factor in the success of IoT applications. It helps ensure that data is reliably sent and received on time and that the data is accurate. QoS also helps ensure that the network is secure so that unauthorized users can't access the data. It is a means of evaluating the quality, efficiency, and performance of IoT devices, systems, and architectures. QoS metrics for IoT applications include reliability, cost, energy consumption, security, availability, and service time [66]. In order for IoT ecosystems to be smarter, they must comply with QoS standards. IoT services and devices must also be defined first in terms of QoS metrics in order to ensure their reliability. Furthermore, users may be able to specify their needs and requirements. For QoS assessment, there are several approaches to choose from, and in [67], quality factors and approaches are traded off. It is, therefore, necessary to deploy good-quality models in order to overcome this trade-off.

4.6 Summary

Researchers and developers around the world have been drawn to IoT advancements in recent years. In order to provide society with as much benefit as possible from IoT, researchers and developers are working together. Smart sustainable cities are increasingly relying on big data applications enabled by the IoT as a form of pervasive computing to improve their operational functioning and planning to contribute to the development of a more environmentally sustainable society. IoT enables smart cities to reduce their energy consumption and carbon emissions while increasing the efficiency of their services. It also improves city operations by helping to monitor traffic, air quality, and water resources, as well as providing better access to data that can help inform decision-making. Integrating IoT devices into a city's infrastructure, it provides real-time data and insights into the energy consumption of buildings, transportation systems, and other public services. This data is used to optimize energy usage and identify opportunities for reducing energy costs while helping to improve the quality of life for citizens.

Cloud, fog, edge, and quantum computing are promising techniques that reduces energy costs in smart cities by optimizing the utilization of energy resources. They also provide real-time insights into energy usage, enabling the management of energy demand and minimizing energy waste. Additionally, they enable predictive maintenance of energy infrastructure, allowing minimized downtime and maintenance costs. Cloud computing help cities become more efficient and sustainable through the use of data analytics and automation. This helps cities reduce their energy consumption, improve air quality and traffic, and provide citizens with more efficient services. Fog computing is a distributed computing model that allows for edge processing, which eliminates the need to send data to the cloud for processing. This reduces the load on the cloud, reduces latency, and reduces the amount of energy needed to process the data. This makes fog computing an ideal solution for sustainable smart cities. Edge computing is a way of processing data closer to the source of the data, which helping reduce energy consumption, power use, and the overall carbon footprint of a smart city. Edge computing also helps improve the efficiency of a smart city by reducing latency and improving response times. Quantum computing enables city planners to develop efficient solutions to problems such as energy usage, traffic congestion, and waste management. It also enables them to build systems that are more resilient to changes in weather and climate.

References

1. S.E. Bibri, *The Shaping of Ambient Intelligence and the Internet of Things: Historico–Epistemic, Socio–Cultural, Politico–Institutional and Eco–Environmental Dimensions* (Springer–Verlag, Berlin/Heidelberg, 2015)
2. S.E. Bibri, J. Krogstie, On the social shaping dimensions of smart sustainable cities: A study in science, technology, and society. Sustain. Cities Soc. **29**, 219–246 (2016)
3. https://www.un.org/development/desa/en/news/population/2018-revision-of-world-urbanization-prospects.html. Accessed on 8 Mar 2023
4. https://www.idc.com/getdoc.jsp?containerId=IDC_P37477. Accessed on 8 Mar 2023
5. https://flex.com/insights/live-smarter-blog/why-5g-will-make-smart-cities-reality. Accessed on 8 Mar 2023
6. https://www.marketsandmarkets.com/Market-Reports/5g-iot-market-14027845.Html?gclid =Cj0KCQiA_JWOBhDRARIsANymNOZ6-EnIIRTTV_VE59CRCwz3o_hEaHxq4fd3aR yoZNlNf1fpb0y0NwsaAjXGEALw_wcB. Accessed on 8 Mar 2023

7. B.N. Silva, M. Khan, K. Han, Internet of things: A comprehensive review of enabling technologies, architecture, and challenges. IETE Tech. Rev. **35**(2), 205–220 (2017)
8. J. Lee, Performance evaluation of IEEE 802.15. 4 for low-rate wireless personal area networks. IEEE Trans. Consum. Electron. **52**(3), 742–749 (2006)
9. X. Ge, S. Tu, G. Mao, C.X. Wang, T. Han, 5G ultra-dense cellular networks. IEEE Wirel. Commun. **23**(1), 72–79 (2016)
10. R. Sanchez-Iborra, M.D. Cano, State of the art in LP-WAN solutions for industrial IoT services. Sensors **16**(5), 708/1–14 (2016)
11. A. Deshpande, C. Guestrin, S.R. Madden, J. M. Hellerstein, W. Hong, Model driven data acquisition in sensor networks, in *Proceedings of the 13th International Conference on Very Large Data Bases*, (vol 30, no 1, 2004), pp. 588–599
12. D. Bandyopadhyay, J. Sen, Internet of things: Applications and challenges in technology and standardization. Wirel. Pers. Commun. **58**, 49–69 (2011)
13. L. Sanchez, L. Munoz, J.A. Galache, P. Sotres, J.R. Santana, V. Gutierrez, et al., Smart Santander: IoT experimentation over a smart city testbed. Comput. Netw. **61**, 217–238 (2014)
14. N. Subramanian, A. Jeyaraj, Recent security challenges in cloud computing. Comput. Electr. Eng. **71**, 28–42 (2018)
15. P. Mell, T. Grance, *The NIST Definition of Cloud Computing* (National Institute of Standards and Technology. U.S. Department of Commerce, Gaithersburg, MD, USA, Technical Representative, NIST Special Publication, 2011)
16. S. Marston, Z. Li, S. Bandyopadhyay, J. Zhang, A. Ghalsasi, Cloud computing-the business perspective. Decis. Support. Syst. **51**(1), 176–189 (2011)
17. D.A. Battleson, B.C. West, J. Kim, B. Ramesh, P.S. Robinson, Achieving dynamic capabilities with cloud computing: An empirical investigation. Eur. J. Inf. Syst. **25**(3), 209–230 (2016)
18. S. Paquette, P.T. Jaeger, S.C. Wilson, Identifying the security risks associated with governmental use of cloud computing. Gov. Inf. Quart. **27**(3), 245–253 (2010)
19. P.-Y. Chen, S.-Y. Wu, The impact and implications of on-demand services on market structure. Inf. Syst. Res. **24**(3), 750–767 (2013)
20. G. Suciu, A. Vulpe, S. Halunga, O. Fratu, G. Todoran, V. Suciu, Smart cities built on resilient cloud computing and secure Internet of Things, in *Proceedings of 19th International Conference on Intelligent Computing, Information and Control Systems*, (May, 2013), pp. 513–518
21. M. Armbrust, A. Fox, R. Griffith, A. D. Joseph, R. Katz, A. Konwinski, … M. Zaharia, A view of cloud computing. Commun. ACM **53**(4), 50–58 (2010)
22. H. Cloud, The nist definition of cloud computing. Natl. Inst. Sci. Technol. (Special Publication) **800**(2011), 1–20 (2011)
23. Y.C. Hu, M. Patel, D. Sabella, N. Sprecher, V. Young, Mobile edge computing—A key technology towards 5G. ETSI White Pap. **11**(11), 1–16 (2015)
24. Y. Mao, C. You, J. Zhang, K. Huang, K.B. Letaief, A survey on mobile edge computing: The communication perspective. IEEE Commun. Surv. Tutorials **19**(4), 2322–2358 (2017)
25. S.E. Bibri, J. Krogstie, ICT of the new wave of computing for sustainable urban forms: Their big data and context-aware augmented typologies and design concepts. Sustain. Cities Soc. **32**, 449–474 (2017)
26. N. Dlodlo, T. Foko, P. Mvelase, S. Mathaba, The state of affairs in internet of things research. Electron. J. Inf. Syst. Eval. **15**(3), 244–258 (2012)
27. A. Ghose, P. Biswas, C. Bhaumik, M. Sharma, A. Pal, A. Jha, Road condition monitoring and alert application: Using in-vehicle Smartphone as Internet-connected sensor, in *Proceedings of the 2012 IEEE 10th International Conference on Pervasive Computing and Communications (PERCOM Workshops)*, (2012), pp. 489–491
28. X. Ren, H. Jiang, Y. Wu, X. Yang, K. Liu, The Internet of Things in the license plate recognition technology application and design, in *Proceedings of the 2012 2nd International Conference on Business Computing and Global Informatization (BCGIN)*, (2012), pp. 969–972

References

29. S. Vongsingthong, S. Smanchat, Internet of things: A review of applications and technologies. Suranaree J. Sci. Technol. **21**(4), 359–374 (2014)
30. J. Shang, Y. Zheng, W. Tong, E. Chang, Inferring gas consumption and pollution emission of vehicles throughout a city, in *Proceedings of the 20th SIGKDD Conference on Knowledge Discovery and Data Mining (KDD 2014)*, (2014), pp. 1027–1036
31. E. Al Nuaimi, H. Al Neyadi, N. Nader, J. Al-Jaroodi, Applications of big data to smart cities. J. Internet Serv. Appl. **6**(25), 1–15 (2015)
32. M. Batty, K.W. Axhausen, F. Giannotti, A. Pozdnoukhov, A. Bazzani, M. Wachowicz, et al., Smart cities of the future. Eur. Phys. J. **214**, 481–518 (2012)
33. M. Ersue, D. Romascanu, J. Schoenwaelder, A. Sehgal, *Management of Networks with Constrained Devices: Use Cases* (IETF Internet, 2014)
34. J. Parello, B. Claise, B. Schoening, J. Quittek, *Energy Management Framework* (IETF internet, 2014)
35. N. Mohamed, J. Al-Jaroodi, Real-time big data analytics: Applications and challenges, in *Proceedings of High Performance Computing & Simulation (HPCS), 2014 International Conference*, (2014), pp. 305–310
36. J. Yin, P. Sharma, I. Gorton, B. Akyoli, Large-scale data challenges in future power grids, in *Proceedings of Service Oriented System Engineering (SOSE), 2013 IEEE 7th International Symposium on IEEE*, (2013), pp. 324–328
37. Information Society Technologies Advisory Group (ISTAG). Working Group on 'ICT and Sustainability (including Energy and Environment)' (2008). http://www.Cordis.lu/ist/istag.htm. Accessed on 20 Mar 2023
38. Y. Zheng, F. Liu, H. Hsieh, U-Air: When urban air quality inference meets big data, in *Proceedings of the 19th ACM SIGKDD International Conference on Knowledge Discovery and Data Mining, ACM*, (2013), pp. 1436–1444
39. S. Li, H. Wang, T. Xu, G. Zhou, Application study on internet of things in environment protection field. Lect. Notes Electr. Eng. **133**, 99–106 (2011)
40. S.H. Lee, J.H. Han, Y.T. Leem, T. Yigitcanlar, Towards ubiquitous city: Concept, planning, and experiences in the Republic of Korea, in *Proceedings of Knowledge—Based Urban Development: Planning and Applications in the Information Era*, (2008), pp. 148–169
41. O. Vermesan, P. Friess, *Internet of Things: Converging Technologies for Smart Environments and Integrated Ecosystems* (River Publishers Series In Communications, London, 2013)
42. A.S. Abdul-Qawy, P.J. Pramod, E. Magesh, T. Srinivasulu, The internet of things (iot): An overview. Int. J. Eng. Res. Appl. **5**(12), 71–82 (2015)
43. J. Gubbi, R. Buyya, S. Marusic, M. Palaniswami, Internet of things (IoT): A vision, architectural elements, and future directions. Futur. Gener. Comput. Syst. **29**(7), 1645–1660 (2013)
44. M. Chui, M. Loffler, R. Roberts, *The Internet of Things* (McKinsey Global Institute, 2010)
45. D. Kyriazis, T. Varvarigou, A. Rossi, D. White, J. Cooper, Sustainable smart city IoT applications: Heat and electricity management and eco conscious cruise control for public transportation, in *Proceedings of the 2013 IEEE 14th International Symposium and Workshops on a World Of Wireless, Mobile and Multimedia Networks (WoWMoM)*, (2014), pp. 1–5
46. X. Fafoutis et al., A residential maintenance-free long-term activity monitoring system for healthcare applications. EURASIP J. Wirel. Commun. Netw. **1**, 1–20 (2016)
47. M.R. Bashir, A.Q. Gill, Towards an IoT big data analytics framework: Smart buildings system, in *Proceedings of IEEE 18th International Conference on High Performance Computing and Communications, IEEE 14th International Conference on smart city, IEEE 2nd International Conference on Data Science and Systems*, (2016), pp. 1325–1332
48. C. Lee, C. Yeung, M. Cheng, Research on IoT based cyber physical system for industrial big data analytics, in *Proceedings 2015 IEEE International Conference on Industrial Engineering and Engineering Management (IEEM)*, (2015), pp. 1855–1859
49. P. Rizwan, K. Suresh, M.R. Babu, Real-time smart traffic management system for smart cities by using internet of things and big data, in *International Conference on Emerging Techno-Logical Trends (ICETT)*, (2016), pp. 1–7

50. C. Vuppalapati, A. Ilapakurti, S. Kedari, The role of big data in creating sense EHR, an integrated approach to create next generation mobile sensor and wear-able data driven electronic health record (EHR), in *Proceedings of 2016 IEEE Second International Conference on Big Data Computing Service and Applications (BigDataService)*, (2016), pp. 293–296
51. D. Mourtzis, E. Vlachou, N. Milas, Industrial big data as a result of IoT adoption in manufacturing. Procedia CIRP **55**, 290–295 (2016)
52. R. Ramakrishnan, L. Gaur, Smart electricity distribution in residential areas: Internet of things (IoT) based advanced metering infrastructure and cloud analytics, in *Proceedings of International Conference on Internet of Things and Applications (IOTA)*, (2016), pp. 46–51
53. M. Mohammadi, A. Al-Fuqaha, S. Sorour, M. Guizani, Deep learning for IoT big data and streaming analytics: A survey. IEEE Commun. Surv. Tutorials **20**(4), 2923–2960 (2018)
54. Z.B. Babovic, V. Protic, V. Milutinovic, Web performance evaluation for internet of things applications. IEEE Access **4**, 6974–6992 (2016)
55. Internet of Things research study: Hewlett Packard Enterprise Report (2015). Available on: http://www8.hp.com/us/en/hp-news/press-release.html?id=1909050#.WPoNH6KxWUk. Accessed on 8 Mar 2023
56. L.D. Xu, W. He, S. Li, Internet of things in industries: A survey. IEEE Trans. Industr. Inform. **10**(4), 2233–2243 (2014)
57. T. Dierks, C. Allen, *The TLS Protocol Version 1.0*, IETF RFC, no. 2246, (1999)
58. M. Pei, N. Cook, M. Yoo, A. Atyeo, H. Tschofenig, *The Open Trust Protocol (OTrP)* (IETF, 2016)
59. R. Roman, P. Najera, J. Lopez, Securing the internet of things. Computer **44**(9), 51–58 (2011)
60. H. Van-der-Veer, A. Wiles, *Achieving Technical, Interoperability-the ETSI Approach*, ETSI White Paper No. 3 (2008). http://www.etsi.org/images/files/ETSIWhitePapers/IOP%20whitepaper%20Edition%203%20final.pdf. Accessed on 8 Mar 2023
61. A. Colacovic, M. Hadzialic, Internet of things (IoT): A review of enabling technologies, challenges and open research issues. Comput. Netw. **144**, 17–39 (2018)
62. M. Noura, M. Atiquazzaman, M. Gaedke, Interoperability in internet of things infrastructure: Classification, challenges and future work, in *Proceedings of Third International Conference, IoTaaS 2017*, (Taichung, Taiwan, 20–22 September, 2017), pp. 11–18
63. S.G. Tzafestad, Ethics and law in the internet of things world. Smart Cities **1**(1), 98–120 (2018)
64. M. Mosko, I. Solis, E. Uzun, C. Wood, CCNx 1.0 protocol architecture, in *A Xerox company, computing science laboratory PARC*, (2017)
65. Y. Wu, J. Li, J. Stankovic, K. Whitehouse, S. Son, K. Kapitanova, Run time assurance of application-level requirements in wireless sensor networks, in *Proceedings of 9th ACM/IEEE International Conference on Information Processing in Sensor Networks*, (Stockholm, Sweden, 21–16 April, 2010), pp. 197–208
66. L. Huo, Z. Wang, Service composition instantiation based on cross-modified artificial bee Colony algorithm. China Commun. **13**(10), 233–244 (2016)
67. G. White, V. Nallur, S. Clarke, Quality of service approaches in IoT: A systematic mapping. J. Syst. Softw. **132**, 186–203 (2017)

Chapter 5
6G-IoT Framework for Sustainable Smart City: Vision and Challenges

5.1 Introduction

A smart city is one in which existing networks and essential services are enhanced via the implementation of digital and communication technology, for the benefit of its residents and industries. While transitioning towards a society of autonomous systems, a high-data-rate network with stable connectivity will be crucial to support applications in various aspects of the city, including industry, health, highways, oceans, and space. Smart cities are supported by a variety of technologies; 5G and beyond technology is one of the most important areas, including 6G technologies, which are still under development. By 2035, 6G is projected to become the predominant access point [1].

Aiming for the implementation of several number of United Nations Sustainable Development Goals (UN SDGs) by 2030, the commercial rollout of the 6G communication system is anticipated. Experts believe that 6G communications will not only enhance global productivity and growth but also establish new business models and bring revolutionize changes to various sectors of society. The SDGs were established by the United Nations as a framework for identifying the possibilities and challenges associated with a desirable future world [2]. There are a variety of topics addressed within the SDGs, including eradicating poverty, gender equality, climate change, and smart cities. 6G technology could benefit the UN SDGs and serve as a catalyst for its advancement by contributing to the use of 6G technology [3–6].

Fifth-generation (5G) wireless networks have already been deployed in several regions of the world but is unable to provide a fully automated and intelligent network that offers everything as a service [7, 8]. Despite the fact that 5G communication systems will provide significant gains over current systems, they will not be able to meet the requirements of future smart and autonomous systems beyond 10 years [9, 10]. The demand for reliable Internet of Things (IoT) systems that deal with increasing demand in various applications is developing drastically in the

© The Author(s), under exclusive license to Springer Nature Switzerland AG 2023
P. Mishra, G. Singh, *Sustainable Smart Cities*,
https://doi.org/10.1007/978-3-031-33354-5_5

advanced communication system. The 6G-IoT vision can be realized through sensors that connect to computers, resulting in cost savings, improved usage, and increased productivity. IoT technology has changed the way we live, with ubiquitous applications such as smart cities, augmented reality (AR), industrial IoT, and more. Because of the tremendous growth in the number of IoT devices, upcoming communication networks aim to achieve great spectral efficiency, low latency, and vast connection.

The number of IoT devices is expected to exceed 25 billion by 2025 [11], making it difficult for present multiple access techniques to cope with such a wide variety of devices. Even 5G communication systems are incapable of supporting huge number of IoT devices. B5G wireless systems need to be established with additional enticing characteristics to attain the objective of 6G and overcoming the restrictions of 5G for enabling future challenges. By bringing new integration of forthcoming services such as ubiquitous sensing intelligence and new human-human and human-machine interaction, 6G communication networks will address the shortcomings of the 5G system. The digital, physical, and human worlds will fluidly merge in the 6G future to trigger extrasensory experiences as depicted in Fig. 5.1. The role of next-generation networks is to bring our physical, digital, and human world experiences together [12]. Future applications will utilize digital worlds as the framework to deliver realistic and high-resolution representations of the physical world, as well as virtual worlds. This paradigm is analogous to how current applications are built upon the multimedia foundation. It entails the extensive use of artificial intelligence (AI) as well as the integration of emerging technologies such as quantum communications, 3D networking, terahertz (THz), intelligent reflecting surface, holographic beamforming, backscatter communication, and proactive caching [13]. The confluence of all previous aspects, such as high reliability, low energy consumption, network densification, high throughput, and huge connection, will be the major drivers of 6G. Similar to earlier generations, 6G systems incorporate new services and technologies such as AI, wearable technologies, implantation, autonomous vehicles, computing reality equipment, sensors, and 3D mapping [14]. The ability to handle huge amounts of data and very high data rate connectivity per devices is the most significant necessity for 6G wireless networks.

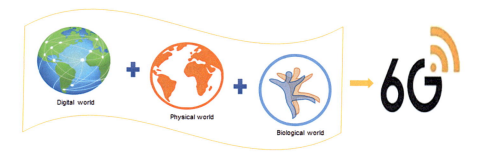

Fig. 5.1 Connection in 6G era integrating digital, physical, and biological world

5.1 Introduction   99

In 6G, super-fast and high-power processors will be used to reduce network latency at both the network and the end device level. Other than that, 6G mobile phones will offer incredible capabilities such as very high accuracy in determining location, positioning, and distance. This will aid communication and positioning in marine and undersea environments. The suggested 6G network needs are super-high data rates, ultra-low latency, increased mobility and coverage, flexible and efficient connectivity, spectral efficiency, and increased network security. The data speeds for 6G are estimated to be in the terabits per second range, with a latency of less than 1 millisecond. This technology will drive the Internet of Everything (IoE), with 107 connections per square kilometer.

5.1.1 Evolution of Generations: 1G to 6G

As we progress toward this next generation of mobile network, let us look at the experiences that each generation brought with it and what the future 6G might hold. Now, 5G promises high-speed experience and low-latency mobile internet services on the go. As we advance towards this next generation of mobile network, let us examine the experiences that each generation has brought with it. The evolution of generations from 1G to 6G is presented in Fig. 5.2.

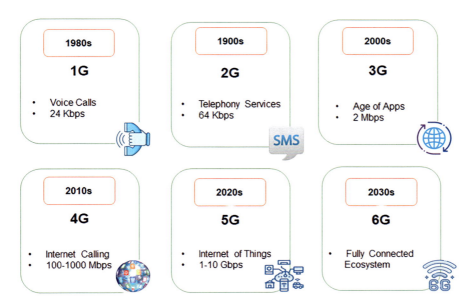

Fig. 5.2 Evolution of generations from 1G to 6G

- **1G: Voice Calls**

It was a time when phones were thick, heavy, and bulky. They had no screens and came with big antennas and massive batteries. The network reception was sketchy, and the on-battery time was abysmal. Nevertheless, this is where the mobile network story started. The first generation enabled the communication between two supported devices using a wireless network. Based on the analog system, 1G supported only voice calls, and those too are of poor quality because of interference. Besides, 1G worked in a fixed geographical area because of lack of roaming support by the network.

- **2G: Telephony Services**

The second generation fixed the issues that marred the first-generation mobile network and introduced new capabilities. The analog system of the first generation was now replaced by a much-advanced digital technology for wireless transmission called the Global System for Mobile Communications (GSM). With digital underpinning, the 2G supported better-quality voice calls and data services such as short message service (SMS) and multimedia messaging service (MMS). Besides, this mobile network enabled roaming facility, allowing users to attend calls and send and receive texts and multimedia content on the go. The 2G network enabled true telephony services. It later received internet support in the form of GPRS (General Packet Radio Service) and EDGE (Enhanced Data rates for GSM Evolution), but that alone wasn't enough for a generational shift. Therefore, there was also 2.5G before the world moved to 3G.

- **3G: Age of Apps**

The third-generation mobile network introduced high-speed internet services, which set the stage for smartphones and app ecosystems. While 3G enabled the concept of mobile television, online radio services, and emails on phones, it is video calling and mobile phone apps that really define the 3G era. This was the time when iPhones and Android smartphones started making inroads. The early iteration of 3G supported internet speed in kilobytes per second (Kbps). Like 2G, there was no direct shift from 3G to 4G. There was a 3.5G, which was earmarked for better internet speeds in megabyte per second (Mbps) with the introduction of technologies such as HSDPA (High-Speed Downlink Packet Access) and HSUPA (High-Speed Uplink Packet Access).

- **4G: Internet Calling**

3G set the base for 4G, which is the generation of mobile network we are currently on. The concepts introduced by 3G such as high-definition voice calls, video calls, and other internet services become a reality in 4G—due to a higher data rate and advanced multimedia services supported by the mobile network. It enhanced the LTE (long-term evolution) system, which significantly improves data rate and allows simultaneous transmission of voice and data. Internet calling, or VoLTE (Voice over LTE), is one of the many advantages of the 4G mobile network. The

5.1 Introduction

network also enables Voice over Wi-Fi (VoWi-Fi), which allows voice calls in areas with low or no network reception.

- **5G: IoT and Enterprises**

From 1G to 4G, each successive generation of communication technology brought about significant changes in the network, enhancing the use cases of the previous generation and introducing new ones. 5G, however, is expected to be a little different, in the sense that it will be another mobile network geared toward not just smartphone users but also enterprises.

This is because the next generation of network would bring not just improvement in data speeds but also latency and throughput. The low latency and high throughput make the network ideal for enterprise use, especially with regard to automation and connected ecosystem. On the consumer side, the network would deliver high internet speeds and would likely play a crucial role in enabling technologies such as the metaverse.

- **6G: Connected Ecosystem**

6G is promoted to drive the adoption of 5G use cases at scale through optimizations and cost reduction, especially at the enterprise level. Take the concept of the metaverse, for example. It is one of the 5G use cases, which promises to disrupt both traditional and digital spaces. With 6G, the metaverse would not just evolve into a final model but is also likely to unify with the physical world with the help of AI and machine learning (ML). This is because the most notable aspect of 6G would be its ability to sense the environment, people, and objects, according to telecom gear maker Nokia Bell Labs. To accomplish so, 6G will use sub-terahertz and terahertz spectrum, i.e., 300 GHz to 10 THz (results in high data rates) [15, 16], which has a higher frequency range than millimeter-wave (mm-Wave) spectrum (30–300 GHz) used in 5G [17]. Optical wireless approaches [18], such as visible light communication (VLC) [19] and free-space optical communication (FSOC) [20], are also extensively studied. In addition, technologies like as reconfigurable intelligent surfaces (RIS) [21–23], cell-free massive MIMO [24], and AI are discussed, all of which are projected to accelerate the implementation of 6G.

There is presently limited information available concerning IoT-based 6G technologies. However, it is expected that international standardization groups would finalize 6G standards by 2030 [25, 26]. In [27], authors suggest on futuristic possibilities for next-generation wireless systems, extending the concept of 5G technology to 6G networks. Currently in the early stages of development, self-driving cars would make life safer and more comfortable. The authors present the predicted technology and prospective uses of 6G in [28–31]. With use cases, vision, and technologies, the articles [32–35] present the system-level view of the 6G scenario. A 6G network enabled by mm-Wave communications in satellite communication is discussed in article [36]. In [37], a comparison of the key performance indicators (KPIs) for 5G and 6G is conducted along with the vision for next-generation wireless communication systems [38]. Blockchain and AI are two technologies that have the potential to revolutionize communication networks in the future [39]. 6G-IoT mobile devices will be able to access terahertz (THz) and mm-Wave frequency bands, enabling new applications and coverage [40]. Distances will be broken through holographic technologies, virtual reality (VR), and augmented reality (AR). In [41], the potential major supporting technologies for 6G are discussed, which include blockchain-assisted network decentralization and ML-based intelligent communication systems. In [42], practical applications such as the IoT, holography, VR, ML, VLC, and autonomous vehicles are discussed. The technologies that will be used in 6G-IoT systems will undergo a substantial transformation. The contributions of this chapter is summarized as follows:

- In a brief discussion of smart cities, we highlight the future of wireless connectivity, mobile data, and possible paths to a 6G communication system. The vision of 6G-IoT for smart city, based on a comparison of the essential requirements with those of the existing generations of wireless communication networks, is addressed.
- The proposed framework for 6G-IoT for a sustainable smart city is presented, highlighting the 6G-IoT networks' key aspects.
- Expected 6G use cases and the role of different 6G enabling technologies such as artificial intelligence/machine learning, spectrum bands, sensing network, extreme connectivity, new network architectures, security, and trust are discussed in detail.
- The challenges, open issues, and research directions required to achieve the goals of 6G-IoT for sustainable smart city are described.

The rest of the chapter is organized as follows. 6G-IoT vision and requirements are presented in Sect. 5.2. The proposed framework for sustainable smart city is presented in Sect. 5.3. Various enabling technologies and use cases for 6G-IoT-based communication system are discussed in Sects. 5.4 and 5.5, respectively. The open research challenges and research directions are discussed in Sect. 5.6. Finally, Sect. 5.7 concludes with the summary.

5.2 Vision and Requirements

In the future, it is likely that sustainable smart cities powered by IoT will become more popular, necessitating the development of next-generation wireless networks. 6G must transform human society into an intelligent society in order to fulfill the wireless transmission expectations of 2030 or beyond [43, 44]. However, 5G has a number of inherent limitations and challenges that have hindered it from meeting its goals. To meet these demands, we need to switch to next generation that supports huge numbers of users. Despite the fact that 5G networks can handle a variety of IoT services, they may not be able to fully meet the needs of emerging applications in terms of requisite internet speed, security, range, and infrastructures in the near future in the context of super-intelligent city [45]. Wireless technologies in 6G are expected to overcome the limitations of 5G networks. The rapid growth of

5.2 Vision and Requirements

IoT-based smart city applications has set the path for existing wireless networks to evolve by connecting people, data, and things through new and advanced technologies in order to provide a wide range of services.

6G-IoT for sustainable smart city envisions intelligent personal edge, physical to cyber fusion, health analytics services, sensing health indicators, super-functional products for multidimensional design technology, printed electronic products, mobility as a service for objects and infrastructure communication, autonomous safety management, human user interface (UI) through AI interfaces, and multi-object tracking. The smart city services involve AI observing and catering services, management of entire city logistics, and sentient safety and comfort. To make it possible, we need techniques for massively scalable systems, to develop communication between the elements and activity. Wireless multi-access connectivity, personalized AI, projection interfaces, ML, cyber security, edge analysis, sensor fusion, blockchain, 3D IoT design, augmented sensing, printed electronics, multidimensional design technology, and functional materials are the essential requirements for 6G-IoT-based sustainable smart city.

With respect to security, it is very high as compared to 5G. Super IoT systems offer ultra-low latency of <1 ms, data rates in the terabits per second range and utilize the THz spectrum. The types of services required for 6G-IoT are MBRLLC (massive machine-type communications with ultra-reliable low-latency communication), mURLLC (mission-critical ultra-reliable low-latency communication), HCS (hybrid context sensing), and MPS (mobile positioning services). MBRLLC is a technology designed to meet stringent requirements for rate reliability, latency, and energy efficiency in mobility environments. mURLLC involves ultra-high reliability, massive connectivity, massive reliability, and scalable URLLC. QoPE (quality of physical experience) encompasses the capture of raw wireless metrics as well human and physics factors, which together constitute the HCS approach. Similarly, MPS offers control stability, computing latency, localization accuracy, sensing and mapping accuracy, latency, and reliability for communication and energy. The network characterizations of 6G are intelligentization, cloudization, softwarization, virtualization, and slicing for self-optimization, self-organization, and self-reconfiguration. The fundamental 6G key technologies for IoT include THz communication, Spatial Modulation Multiple-Input Multiple-Output (SM-MIMO), Large Intelligent Surface (LIS) and Hybrid Beamforming (HBF), Orbital Angular Momentum (OAM) multiplexing, laser and VLC, blockchain-based spectrum sharing, quantum communication and computing, and AI/ML key technologies. The key applications propelling futuristic use cases of 6G include holographic verticals and society, tactile/haptic internet, full sensory digital sensing and reality, fully automated driving, Industrial IoT, space travel, deep-sea sightseeing, and Internet of Bio-Nano Things.

In addition to smart applications, 5G networks will support a wide range of IoT-based services. The upcoming services, like haptics, Extended reality (XR)/ Virtual reality (VR)/ Mixed reality (MR), Brain Computer Interface (BCI), etc., are built on extremely high levels of reliability, mobility, and data rates. According to the vision [46], 6G will be enhanced to acquire 10–100 times higher data rates, higher capacity, higher efficiency, reduced delay, and wide coverage. The IoE, as well as innovative technologies such as AI, blockchain, edge computing, cloud computing, and so on, will enable 6G to support a ubiquitous intelligent mobile society. The 6G network is personalized and intelligent and can be realized via virtualized personal mobile communications combined with AI. The AI technologies allow for automatic network maintenance, efficient network administration, flexible network management, and smart network recognition. High frequencies such as mm-Wave, terahertz, or visible light will be used in 6G. Furthermore, 6G can utilize flexible frequency sharing methods to boost frequency reuse efficiency. 6G will implement a security system that is indigenous or uses an integrated security function. It can be used on land, air, and sea, with smart mobility management technologies, creating a global ubiquitous mobile broadband communication system. The 6G network should integrate terrestrial, satellite, short-range device-to-device communications, and so on, for broader and deeper coverage area. As the current work suggests, the 6G IoT will become more intelligent and achieve the system targets. In 6G IoT applications that generate enormous data, AI technologies such as ML train models based on big data in order to achieve satisfactory performance. AI can be executed on a device, at the network edge, or even in the cloud depending on the application.

Additionally, multilevel AI is used to perform deep data mining and develop intelligent models. The 6G system has reached significant advancements on performance in terms of mobility support (1000 km/h), frequency (sub-6 GHz, mm-Wave for mobile access, THz bands (above 300 GHz)), latency (0.1 ms), data rate ($\geq$ 1 Tb/s), network density (10^7 nodes/km^2), processing delay (10 ns), and other KPIs, with haptic communication, autonomous vehicle, AI, and satellite integration. Compared with previous generations, the 6G system has reached significant advancements on performance in terms of mobility, latency, data rate, reliability, and other KPIs, as shown in Table 5.1.

5.3 6G-IoT Framework for Sustainable Smart City

We presented a simplified six-layer 6G-IoT smart city structure in this section, as shown in Fig. 5.3. It is comprised of six interconnected layers that allow for secure data transfer with high data throughput, increased efficiency, wide coverage, and low latency. This process involves data collection, processing, analysis, and exchange of information between devices and communication networks [47–49]. Using 6G technology, each layer is described as follows:

5.3.1 Sensing Layer

The sensing layer consists of physical components, such as smart sensors, actuators, and mobile units. The main function of this layer is to detect and continuously communicate information to the network layer.

5.3 6G-IoT Framework for Sustainable Smart City

Table 5.1 Requirements and KPIs of 6G

Sr. no.	KPIs/ requirements	5G	6G
1.	Peak data rate	≥ 10–20 Gb/s	≥ 1 Tb/s
2.	Frequency	Sub-6 GHz, mm-Wave for fixed access	Sub-6 GHz, mm-Wave for mobile access, THz bands (above 300 GHz)
3.	Latency	1 ms	0.1 ms
4.	Mobility support	500 km/h	1000 km/h
5.	Average data rate	100 mbps	1 Gbps
6.	Spectrum efficiency	3–5 × that to 4G	5 × that to 5G
7.	Network density	10^6 nodes/km^2	10^7 nodes/km^2
8.	Processing delay	100 ns	10 ns
9.	Service type	eMBB, mMTC, and URLLC	MBRLLC, mURLLC, HCS, MPS
10.	Architecture	Dense small cells, mm-Wave small cells of about 100 m (fixed access)	Cell-free (for mobile and fixed access), tiny THz cells
11.	Haptic communication	Partial	Yes
12.	Autonomous vehicle	Partial	Yes
13.	Artificial intelligence	Partial	Yes
14.	Satellite integration	No	Yes

5.3.2 Network Layer

The network layer, also known as the internet layer, comprises Wi-Fi, beyond 5G/6G networks, and gateways for communication. This layer enables sensor data to be transmitted and processed between physical devices and servers. 3D networking, nano-networking, bio-networking, and optical networking are some of the new networking technologies for 6G [50]. Future 6G smart services for smart city applications are expected to include B-IoT (biological cell communication) and N-IoT (molecular communication).

5.3.3 Communication Layer

The communication layer serves as the foundation for 6G-IoT framework because it transfers all the information/data within the layers. A 6G-IoT system will enable a variety of smart applications by utilizing novel communication technologies. 3D wireless communication, quantum communication, THz communication, VLC, and

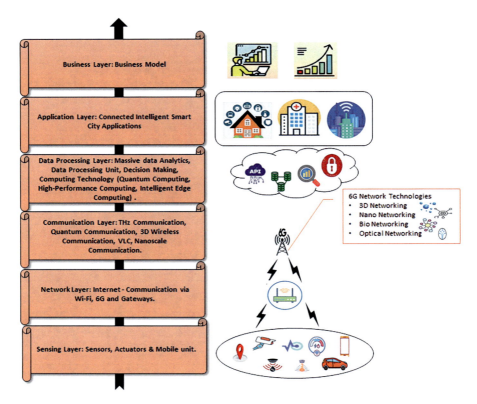

Fig. 5.3 6G-IoT framework for sustainable smart city

nanoscale communication are examples of these communication technologies [50]. In 3D communication, airborne and ground networks are integrated. An unmanned aerial vehicle (UAV) and a low-orbit satellite can work as a base station for 3D communications. In 6G-IoT network, quantum communication offers high levels of security. In addition to millimeter-wave bands, 6G uses THz wavelengths to enable high data throughput. THz communication typically employs frequencies ranging from 0.1 to 10 THz. High security, short range, low-power consumption, and robustness are its distinguishing features. Using a visible light spectrum ranging from 430 THz to 790 THz, VLC can be used to enable a variety of 6G-IoT-based applications. In addition, VLC provides substantial bandwidth and interference-free communication compared to radio-frequency waves. Nanoscale communication uses an incredibly small wavelength and is capable of communicating over distances of 1 meter or centimeter.

5.3.4 Data Processing Layer

The data processing layer enables for the storage and analysis of massive amounts of network data. Data and information from edge computing are kept in the cloud in order to achieve final data. The main functions of this layer include massive data storage, data processing, massive data analytics services, decision-making analyzer, computing, and more. 6G-IoT involves various sources of data from multiple smart applications that generate a massive amount of data. Various computing 6G technologies are quantum computing, high-performance computing, and intelligent edge computing. Quantum computing is likely to enhance the area of computing by providing users access to faster speed than they have ever experienced before. High-performance computing is preferred for secure 6G-IoT smart applications due to its ultra-high speed and security. Apart from quantum computing and high-performance computing, 6G will require intelligent edge computing to deliver intelligent on-demand computation as well as on-demand storage capabilities along with incredibly low latency.

5.3.5 Application Layer

6G-IoT applications are able to enable smart cities, smart homes, smart industries, and smart transportation. All devices, sensors, and data are connected to the internet through this layer via wireless 6G connectivity.

5.3.6 Business Layer

The business layer is where the business model is defined. This layer acts as a manager for the 6G-IoT system, providing a platform and allowing smart city applications to be controlled. The profit and business models of the 6G-IoT system are also dealt with by the business layer. At this layer, a variety of data analytics and data science methods are used to gain the required profit or services from smart 6G-enabled IoT devices. It also guarantees that information is secure and used in the best possible way for better services.

Smart city applications are likely to drive additional development in 6G-IoT-oriented cities as the number of smart IoT devices rises. The 6G-IoT framework for sustainable smart cities intends to give city users limitless internet connectivity by meeting their large number of resident's needs. A wide range of data, mobility, scalability, and low latency are supported with high security, wide coverage, minimal implementation costs, and simple and quick troubleshooting. With the proposed framework, data transmission is secured at high data rates and with high performance for a variety of smart city applications, such as smart home systems, smart transportation, smart factories, etc.

5.4 6G-IoT Enabling Technologies

According to Nokia Bell Labs, 6G will be defined by six technological domains which are artificial intelligence and machine learning, spectrum bands, sensing network, extreme connectivity, new network architectures, and security and trust as shown in Fig. 5.4.

5.4.1 Artificial Intelligence and Machine Learning

In the past decade, AI/ML techniques, particularly deep learning (DL) have grown significantly, and they have already been applied in a variety of fields involving image identification and computer vision and tend to range from social networks to security [51, 52]. With the techniques in 5G Advanced, AI/ML will indeed be brought to several parts of the network at many layers and in many functionalities, allowing these technologies to reach their full potential. From beamforming optimization in the radio layers to self-optimizing networks for cell site scheduling, AI/

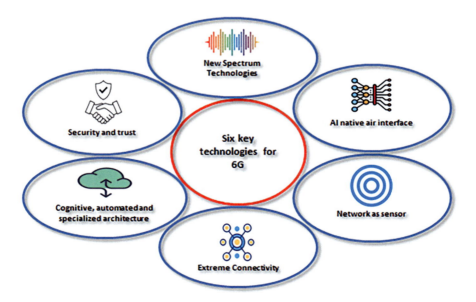

Fig. 5.4 6G-IoT communication system enabling technologies

ML is being used to provide higher efficiency with less complexity. By minimizing the complexity and allowing AI/ML to work out how to best communicate between two endpoints, Nokia Bell Labs anticipates AI/ML to progress from an upgrade to a foundation in 6G.

5.4.2 New Spectrum Technologies

In order to provide radio communication, spectrum is essential. In order to fully exploit a new technology, a new mobile generation requires some new pioneer spectrum. Specifically, the new 6G pioneer spectrum blocks will be in the mid-band 7–20 GHz for urban outdoor cells and low band 460–694 MHz for maximum coverage and above 100 Gbps peak data rates. It is expected that advanced 5G will contribute to better positioning accuracy to the centimeter level, especially for indoor and underground locations where satellite signals are not available, but 6G will extend localization to the THz range using wide spectrum and new spectral ranges.

5.4.3 A Network That Can Sense

The capability of 6G to perceive the surroundings, humans, and object is its most prominent feature. The networks become a stream of ambient data, collecting signals reflecting off obstacles and detecting their type and structure, spatial position, speed, and possibly even material qualities. Digital twin is mainly applied for this purpose, and by merging this data with AI/ML, the network will gain new perspectives from the physical world, making it much more cognitive.

5.4.4 Extreme Connectivity

The URLLC service, which emerged with 5G, will be enhanced and improved in 6G to meet the most stringent connectivity standards, including sub-millisecond latency. Simultaneous transmissions, numerous wireless hops, device-to-device connectivity, and AI/ML could improve network efficiency. In the 6G era, we can expect use cases that involve sub-networks, creating networks of networks where networks serves as the endpoints, creating networks of networks. As an example, a car or body area network may contain hundreds of sensors spread across a small area of less than 100 meters. For the machine system to operate, these sensors must communicate within 100 microseconds with extreme high reliability. Making networks in automobiles and robots entirely wireless will reach new levels for device designers, since they will no longer be required to build long, heavy cable systems.

5.4.5 New Network Architectures

5G is the first solution meant to replace wired communication in the industrial sector. In order to support increased flexibility and specialization, network architectures will need to evolve further to accommodate growing demands and stresses. Emerging connectivity and network parameter solutions that take advantage of AI/ML breakthroughs will result in a new phase of communication networks by lowering operating costs in 6G.

5.4.6 Security and Trust

Cyberattacks are growing more common on all network topologies. The ever-changing nature of the threats necessitates the deployment of robust security mechanisms. 6G networks will be built to withstand attacks such as jamming. When creating new mashed worlds that combine visualizations of actual and virtual objects, privacy concerns will need to be considered.

5.5 6G-IoT Use Cases

The prospects of 6G architecture is envisioned with the utilization of autonomous systems, robots, haptic communication, and IoE. It is characterized under ubiquitous mobile ultra-broadband (uMUB), ultra-high-speed with low-latency communications (uHSLLC), and ultra-high data density (uHDD) services. The implementation of connected robotics and autonomous systems will be aided by 6G technologies. It encourages the practical implementation of self-driving vehicles, which are equipped with a variety of sensors such as sonar, GPS, radar, ranging, light detection, and more. Military, business, law enforcement, delivery service, security, satellite imaging, disaster management, and agriculture, all benefit from 6G networks, which facilitate communications between the UAV and the ground. Furthermore, in the absence of base station, the UAV will be used to facilitate wireless broadcasting and high rate communications. With the utilization of the sense of touch, 6G wireless communication will facilitate haptic communication, benefiting various areas such as robotics, AI sensors, tactic learning for physically challenged individuals, surgical procedures and interactive gaming. Because it enables error-free data transport with no data loss between transmission and reception, 6G will deliver full automation and network integrity based on AI. The 6G communication systems will use the neurological process of sensory integration to transfer data from the five senses. This application benefits greatly from wireless BCI technology [53, 54]. It's a direct line of communication between the brain and the external device. It collects brain signals that are transmitted to a digital device and, eventually, interprets the results.

5.5 6G-IoT Use Cases

Using the infrastructure of the internet, the IoE refers to the seamless integration and autonomous coordination of a rapidly expanding number of computation units and sensors, items, devices, people, and data [55]. It is a form of IoT, which combines four elements in one frame: data, physical devices, people, and process [56]. As opposed to IoT, IoE adds network intelligence to integrate people, data, processes, and physical objects into a single system. Super-smart cities, such as smart automobiles, smart health, and smart industries, rely on the IoT. In this section, we summarize possible 6G communication system including smart home/city, smart industry, satellite terrestrial network, marine and underwater communication, healthcare systems, and extended reality as shown in Fig. 5.5.

5.5.1 Smart City/Home

The environmental and surveillance monitoring, house robotics, and entertainment are all examples of smart home services. In 6G-IoT-based smart city, various application services are involved, including smart utilities such as waste management, water quality monitoring, gas supply, and other city related services. In smart traffic, camera-based traffic monitoring, GPS navigation, and overcrowding monitoring have been implemented in numerous cities across the world. In a smart environment, there are a number of environmental monitoring aspects, such as weather

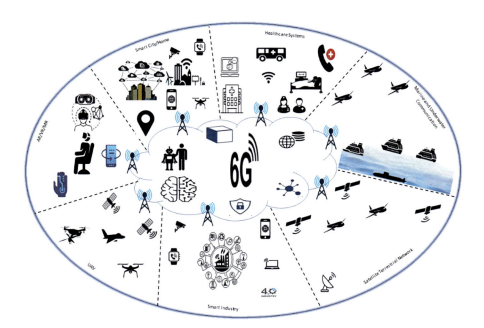

Fig. 5.5 6G-IoT communication system use cases

alert, air pollution indicator, temperature monitoring, sound regulation, and more. With zero latency transmission lines and smart algorithms, self-driving and vehicle-to-vehicle communication applications are rapidly developing. Such services can make city traffic more pleasant and reduce traffic congestion and pollution. However, the building's security may be maintained by using the monitoring and alarm systems. The enhanced qualities of 6G will hasten the development of a super-smart city, resulting in improved living standards, environmental sensing, and automation through AI-based machine-to-machine communication. Through the deployment of smart mobile devices, driverless vehicles, and other technologies, 6G wireless connectivity will make our civilization super-smart. Furthermore, flying cars based on 6G wireless technology will really be deployed in various cities around the globe. Every device is controlled remotely by a smart device, making smart homes a reality.

5.5.2 Smart Healthcare

The 6G network will allow medical systems to have an efficient remote monitoring system. To manage the operation, the smart healthcare system needs extensive, aggregated data, comprising medicinal, administration, and financial information. Multiple types of wearable medical sensors are required for the patient monitor [57, 58]. Additionally, 6G wireless networks are expected to benefit healthcare systems since AR/VR, holographic telepresence, mobile edge computing, and AI will encourage the creation of smart healthcare systems. Using 6G technology, perhaps remote surgeries will indeed be possible. The 6G network, with its high data speeds, reduced latency, and superior capabilities, will facilitate the instant and secure transmission of massive volumes of health information. Biomedical communication is also an important future promise for 6G system [59].

5.5.3 Satellite Terrestrial Network

A satellite terrestrial integrated communication strategy has been considered in the development of 6G-IoT systems to meet a higher demand for greater flexibility and range. Instead of using a regular base station, the user can communicate to a specialized terrestrial satellite terminal. Recent research aims at using UAVs to transmit data to improve coverage, cognitive access, and satellite constellation. Due to the integration of terrestrial and satellite communication, link enables frequency varying from very high frequency (VHF) to Ka band.

5.5.4 Marine and Underwater Communication

Due to the severe attenuation caused by the marine surroundings, the radio-frequency (RF) connections between crewless vehicles, undersea sensor, and research ships suffer from limiting transmission distance. For the very first time, underwater communication is intended to be covered by the 6G era covering tsunami prediction, animal care tracking, and shipwreck inspection.

5.5.5 Extended Reality (XR)

A feature of 6G communication systems is the availability of XR services, such as AR, MR, and VR. The technology of VR provides a computer-simulated 3-dimensional experience in which headsets are used to create realistic feelings and make an imaginary scenario. MR stands for hybrid reality, which combines the actual and virtual worlds to build new environments and visualizations that can be interacted with in real time. Artificial and real-world information communicate to each other in real time, which is a key feature of MR. Computer technology and wearables have integrated actual and virtual surroundings, as well as human-machine communication, to create XR. It combines AR, VR, and MR into a single phrase. XR requires very fast data rate, very low latency, and reliable connectivity, which are all features of the 6G system. Wearable gadgets with advanced characteristics such as XR, high-definition visuals and holograms, and the ability to engage all five senses in communications make it easier to facilitate human-to-human and human-to-thing communications. Three-dimensional objects and AI are the keys to all of these features. Ultra-high-speed with low-latency communications, a 6G service, will make it possible to successfully deploy XR application in the future.

5.6 Open Research Challenges and Research Directions

There are a number of challenges faced by 6G, including capacity, security, coverage, data rate, speed, channel estimation, security, energy efficiency, and spectrum efficiency [60, 61]. Implementation of a satellite mobile communication system is required to achieve ultra-high speed and wider coverage at low cost. The 6G system capacity is enhanced by increasing spectrum bandwidth, spectrum efficiency, efficient modulation, spectrum reuse, or new spectrum utilization methods. Similarly, data rate is increased by increasing spectrum bandwidth and spectrum efficiency. Also, the challenge with high-speed applications like XR, autonomous systems, etc. lies in having energy-efficient hardware at the mobile unit. As a result, further research is required to develop hardware that is compatible with 6G. It is expected that these devices will be energy-efficient with long battery life. Another issue that

Fig. 5.6 Challenges, open issues, and research direction for 6G

6G is expected to address is security and trust in wireless networks [62]. Quantum communication not only improves security but also enhances computational efficiency but needs to be researched more. Within the next couple of years, it is expected that quantum computing will be commercially available and will impose a huge threat on the current cryptographic schemes. As stated in the current state of the art, quantum computing is envisioned to use in 6G communication networks for the detection, mitigation, and prevention of security vulnerabilities. Quantum computing-assisted communication is a novel research area that investigates the possibilities of replacing quantum channels with noiseless classical communication channels to achieve extremely high reliability in 6G. The issues of access control and authentication in blockchain must be solved. For channel estimation in 6G, spatial correlation, low-complexity channel estimation, comprehensive sensing, and deep learning can be applied. To reduce carbon emissions, sustainable and efficient energy harvesting sources can be investigated, yet 6G networks demand more power. As a result, research is needed to lower the cost of softwarization in 6G networks. These challenges, open issues, and possible solution for 6G are shown in Fig. 5.6.

5.7 Summary

By 2030, 5G will not be able to fully serve the growing need for IoT-based sustainable smart cities. As a result, 6G must be implemented to meet the growing demands. We glanced at several features of 6G-IoT wireless networks from different perspectives in this research for assisting in the development of a capable, adaptable, dependable, and secure system. Hence, a conceptual framework of 6G-IoT-based sustainable smart city is presented and discussed in detail. The 6G-IoT will have an impact on health systems, transport, safety, secrecy, and many other areas. The development of sustainable smart cities will be aided by all of these factors. We focused on the vision for 6G communications as well as methods for achieving the aim of 6G-IoT communication in this chapter. Further, we presented a possible 6G network architecture, requirements, and key enabling technologies to be deployed for 6G communication. Finally, we discussed potential challenges and research areas for 6G-IoT systems in order to meet the goal.

References

1. S. Dang, O. Amin, B. Shihada, M.S. Alouini, What should 6G be? Nat. Electron. **3**(1), 20–29 (2020)
2. United Nations (UN) #Envision2030 Sustainable Development Goals, "United Nations (UN)." [Online]. Available: https://sdgs.un.org/goals. Accessed on 8 Mar 2023
3. M. Matinmikko-Blue, S. Aalto, M.I. Asghar, H. Berndt, Y. Chen, S. Dixit, R. Jurva, P. Karppinen, M. Kekkonen, M. Kinnula, P. Kostakos, White paper on 6G drivers and the UN SDGs. arXiv preprint arXiv:2004.14695 (2020)
4. C. de Alwis, A. Kalla, Q.V. Pham, P. Kumar, K. Dev, W.J. Hwang, M. Liyanage, Survey on 6G Frontiers: Trends, applications, requirements, technologies and future research. IEEE Open J. Commun. Soc. **2**, 836–886 (2021)
5. G. Gui, M. Liu, F. Tang, N. Kato, F. Adachi, 6G: Opening new horizons for integration of comfort, security and intelligence. IEEE Wirel. Commun. **27**(5), 126–132 (2020)
6. Z. Zhang, Y. Xiao, Z. Ma, M. Xiao, Z. Ding, X. Lei, G.K. Karagiannidis, P. Fan, 6G wireless networks: Vision, requirements, architecture, and key technologies. IEEE Veh. Technol. Mag. **14**(3), 28–41 (2019)
7. S.J. Nawaz, S.K. Sharma, S. Wyne, M.N. Patwary, M. Asaduzzaman, Quantum machine learning for 6G communication networks: State-of-the-art and vision for the future. IEEE Access **7**, 46317–46350 (2019)
8. S. Wijethilaka, M. Liyanage, Survey on network slicing for internet of things realization in 5g networks. IEEE Commun. Surv. Tutorials **23**(2), 357–994 (2021)
9. L. Bariah, L. Mohjazi, S. Muhaidat, P.C. Sofotasios, G.K. Kurt, H. Yanikomeroglu, O.A. Dobre, A prospective look: Key enabling technologies, applications and open research topics in 6G networks. IEEE Access **8**, 174792–117482 (2020)
10. M. Liyanage, I. Ahmad, A.B. Abro, A. Gurtov, M. Ylianttila, *A Comprehensive Guide to 5G Security* (Wiley Online Library, 2018)
11. S. Mumtaz et al., Terahertz communication for vehicular networks. IEEE Trans. Veh. Technol. **66**(7), 5617–5625 (2017)
12. https://www.nokia.com/about-us/newsroom/articles/6g-explained/. Accessed on 8 Mar 2023
13. E. Calvanese Strinati, S. Barbarossa, J.L. Gonzalez-Jimenez, D. Ktenas, N. Cassiau, L. Maret, C. Dehos, 6G: The next frontier: From holographic messaging to artificial intelligence using subterahertz and visible light communication. IEEE Veh. Technol. Mag. **4**(3), 42–50 (2019)
14. W. Saad, M. Bennis, M. Chen, A vision of 6G wireless systems: Applications, trends, technologies, and open research problems. IEEE Netw. **23**(2), 957–994 (2019)
15. C. Han, Y. Wu, Z. Chen, X. Wang, Terahertz communications (TeraCom): Challenges and impact on 6G wireless systems. arXiv, arXiv:1912.06040 (2019)
16. T.S. Rappaport, Y. Xing, O. Kanhere, S. Ju, A. Madanayake, S. Mandal, A. Alkhateeb, G.C. Trichopoulos, Wireless communications and applications above 100 GHz: Opportunities and challenges for 6g and beyond. IEEE Access **7**, 78729–78757 (2019)
17. W. Hong, Z.H. Jiang, C. Yu, D. Hou, H. Wang, C. Guo, Y. Hu, L. Kuai, Y. Yu, Z. Jiang, et al., The role of millimeter-wave technologies in 5G/6G Wireless communications. IEEE J. Microwav **1**, 101–122 (2021)
18. S. Arai, M. Kinoshita, T. Yamazato, Optical wireless communication: A candidate 6G technology? IEICE Trans. Fundam. Electron. Commun. Comput. Sci. **104**(1), 227–234 (2021)
19. L.U. Khan, Visible light communication: Applications, architecture, standardization and research challenges. Digit. Commun. Netw. **3**(2), 78–88 (2017)
20. S. Chaudhary, A. Amphawan, The role and challenges of free-space optical systems. J. Opt. Commun. **35**(4), 327–334 (2014)
21. M.A.S. Sejan, M.H. Rahman, B.S. Shin, J.H. Oh, Y.H. You, H.K. Song, Machine learning for intelligent-reflecting-surface-based wireless communication towards 6G: A review. Sensors **22**(14), 5405/1–21 (2022)

22. E. Basar, M. Di Renzo, J. De Rosny, M. Debbah, M.S Alouini, R. Zhang, "wireless communications through reconfigurable intelligent surfaces". IEEE Access **7**, 116753–116773 (2019)
23. M. Toumi, A. Aijaz, System performance insights into design of RIS-assisted smart radio environments for 6G, in *Proceedings of 2021 IEEE Wireless Communications and Networking Conference (WCNC)*, (IEEE, 2021), pp. 1–6
24. S. Hu, F. Rusek, O. Edfors, Beyond massive MIMO: The potential of data transmission with large intelligent surfaces. IEEE Trans. Signal Process. **66**, 2746–2758 (2018)
25. A. Mourad, R. Yang, P.H. Lehne, A. De La Oliva, A baseline roadmap for advanced wireless research beyond 5G. Electronics **9**(2), 351 (2020)
26. A. Pouttu, F. Burkhardt, C. Patachia, L. Mendes, G. R. Brazil, S. Pirttikangas, E. Jou, P. Kuvaja, F. T. Finland, M. Heikkila et al., *6g White Paper on Validation and Trials for Verticals Towards 2030s* [Online]. Available: https://www.6gflagship.com/6g-white-paper-on-validation-and-trials-for-verticals-towards-2030s/. Accessed on 8 Mar 2023
27. H. Viswanathan, P.E. Mogensen, Communications in the 6G era. IEEE Access **8**, 57063–57074 (2020)
28. M.Z. Chowdhury, M. Shahjalal, S. Ahmed, Y.M. Jang, 6G wireless communication systems: Applications, requirements, technologies, challenges, and research directions. IEEE Open J. Commun. Soc. **1**, 957–975 (2020)
29. K. Lin et al., Green video transmission in the mobile cloud networks. IEEE Trans. Circuits Syst. Video Technol. **27**(1), 159–169 (2017)
30. K. Sheth, K. Patel, H. Shah, S. Tanwar, R. Gupta, N. Kumar, A taxonomy of AI techniques for 6G communication networks. Comput. Commun. **161**, 279–303 (2020)
31. N. Kato, B. Mao, F. Tang, Y. Kawamoto, J. Liu, Ten challenges in advancing machine learning technologies toward 6G. IEEE Wirel. Commun. **27**(3), 99–103 (2020)
32. M. Giordani, M. Polese, M. Mezzavilla, S. Rangan, M. Zorzi, Toward 6G networks: Use cases and technologies. IEEE Commun. Mag. **58**(3), 55–61 (2020)
33. P. Yang, Y. Xiao, M. Xiao, S. Li, 6G wireless communications: Vision and potential techniques. IEEE Netw. **33**(4), 70–75 (2019)
34. Y. Lu, X. Zheng, 6G: A survey on technologies, scenarios, challenges, and the related issues. J. Ind. Inf. Integr. **19**, 100158/1–14 (2020)
35. M. Giordani, M. Zorzi, Satellite communication at millimeter waves: a key enabler of the 6G era, in *Proceedings of 2020 International Conference on Computing, Networking and Communications (ICNC)*, (IEEE, 2020) pp. 383–388
36. Y. Zhou, L. Liu, L. Wang, N. Hui, X. Cui, J. Wu, Y. Peng, Y. Qi, C. Xing, Service aware 6G: An intelligent and open network based on convergence of communication, computing and caching. Digit. Commun. Netw **6**(3), 253–260 (2020)
37. F. Tariq, M.R. Khandaker, K.K. Wong, M.A. Imran, M. Bennis, M. Debbah, A speculative study on 6G. IEEE Wirel. Commun. **27**(4), 118–125 (2020)
38. Z. Yajun, Y. Guanghui, X. Hanqing, 6G mobile communication networks: Vision, challenges, and key technologies. Sci. Sin. Inf. **49**(8), 963–987 (2019)
39. F. Jameel, U. Javaid, B. Sikdar, I. Khan, G. Mastorakis, C.X. Mavromoustakis, Optimizing blockchain networks with artificial intelligence: Towards efficient and reliable IoT applications, in *Proceedings of Convergence of Artificial Intelligence and the Internet of Things*, (Springer, 2020), pp. 299–321
40. A.A. Raja, M.A. Jamshed, H. Pervaiz, S.A. Hassan, Performance analysis of UAV-assisted backhaul solutions in THz enabled hybrid heterogeneous network, in *Proceedings of IEEE INFOCOM 2020 – IEEE Conference on Computer Communications Workshops (INFOCOM WKSHPS)*, (2020), pp. 628–633
41. A. Yastrebova, R. Kirichek, Y. Koucheryavy, A. Borodin, A. Koucheryavy, Future networks 2030: Architecture & requirements, in *Proceedings of 10th International Congress on Ultra-Modern Telecommunications and Control Systems and Workshops (ICUMT)*, (2018), pp. 1–8
42. D.P.M. Osorio, I. Ahmad, J.D.V. Sanchez, A. Gurtov, J. Scholliers, M. Kutila, P. Porambage, Towards 6G-enabled internet of vehicles: Security and privacy. IEEE Open J. Commun. Soc. **3**, 82–105 (2022)

References

43. S. Chen, Y.-C. Liang, S. Sun, S. Kang, W. Cheng, M. Peng, Vision, requirements, and technology trend of 6G: How to tackle the challenges of system coverage, capacity, user data-rate and movement speed. IEEE Wirel. Commun. **27**(2), 218–228 (2020)
44. V. Ziegler, S. Yrjola, 6G indicators of value and performance, in *Proceedings of 2nd 6G Wireless Summit (6G SUMMIT)*, (IEEE, 2020), pp. 1–5
45. R. Khan, P. Kumar, D.N.K. Jayakody, M. Liyanage, A survey on security and privacy of 5G technologies: Potential solutions, recent advancements, and future directions. IEEE Commun. Surv. Tutorials **22**(1), 196–248 (2019)
46. X. You, C.-X. Wang, J. Huang, X. Gao, Z. Zhang, M. Wang, Y. Huang, C. Zhang, Y. Jiang, J. Wang, et al., Towards 6G wireless communication networks: Vision, enabling technologies, and new paradigm shifts. SCIENCE CHINA Inf. Sci. **64**(1), 1–74 (2021)
47. P.K. Padhi, F. Charrua-Santos, 6G enabled industrial internet of everything: Towards a theoretical framework. Appl. Syst. Innov. **4**(1), 11/1–28 (2021)
48. V. Ziegler, H. Viswanathan, H. Flinck, M. Hoffmann, V. Raisanen, K. Hatonen, 6G architecture to connect the worlds. IEEE Access **8**, 173508–173520 (2020)
49. T. Huang, W. Yang, J. Wu, J. Ma, X. Zhang, D. Zhang, A survey on green 6G network: Architecture and technologies. IEEE Access **7**, 175758–175768 (2019)
50. L.U. Khan, I. Yaqoob, M. Imran, Z. Han, C.S. Hong, 6G wireless systems: A vision, architectural elements, and future directions. IEEE Access **8**, 147029–147044 (2020)
51. B. Schneier, Artificial intelligence and the attack/defence balance. IEEE Secur. Priv. **16**(2), 96–96 (2018)
52. Y. Sun, J. Liu, J. Wang, Y. Cao, N. Kato, When machine learning meets privacy in 6G: A survey. IEEE Commun. Surv. Tutorials **22**(4), 2694–2724 (2020)
53. A.K. Tripathy, S. Chinara, M. Sarkar, An application of wireless brain–computer interface for drowsiness detection. Biocybern. Biomed. Eng. **36**(1), 276–284 (2016)
54. S.R.A. Jafri et al., Wireless brain computer interface for smart home and medical system. Wirel. Pers. Commun. **106**(4), 2163–2177 (2019)
55. Internet of Everything. [Online]. Available: https://ioe.org/ (2019). Accessed on 8 Mar 2023
56. CISCO, *The Internet of Everything*. [Online]. Available: https://www.cisco.com/c/dam/en_us/about/business-insights/docs/ioe-value-at-stake-public-sector-analysis-faq.pdf (2019). Accessed on 8 Mar 2023
57. E. Yaacoub, K. Abualsaud, T. Khattab, A. Chehab, Secure transmission of IoT mhealth patient monitoring data from remote areas using DTN. IEEE Netw. **34**(5), 226–231 (2020)
58. C. Xie, P. Yang, Y. Yang, Open knowledge accessing method in IoT-based hospital information system for medical record enrichment. IEEE Access **6**, 15202–15211 (2018)
59. R. Lin et al., Wireless battery-free body sensor networks using near-field-enabled clothing. Nat. Commun. **11**(444), 1–10 (2020)
60. P. Porambage, G. Gur, D.P.M. Osorio, M. Liyanage, M. Ylianttila, 6G Security challenges and potential solutions, in *Proceedings for 2021 Joint European Conference on Networks and Communications (EuCNC) and 6G Summit, IEEE*, (2021), pp. 1–6
61. M. Latva-Aho, K. Leppanen, *Key Drivers and Research Challenges for 6G Ubiquitous Wireless Intelligence*, White Paper (6G Flagship, Oulu, 2019)
62. M. Wang, T. Zhu, T. Zhang, J. Zhang, S. Yu, W. Zhou, Security and privacy in 6G networks: New areas and new challenges. Digit. Commun. Netw. **6**(3), 281–291 (2020)

Chapter 6
Artificial Intelligence for Sustainable Smart Cities

6.1 Introduction

Artificial intelligence (AI) seems to be permeating our lives in many different and interesting ways. It is becoming more critical in many different industries, like healthcare, entertainment, financial services, and education, owing to its capacity to effectively tackle complex challenges. AI contributes to the convenience and efficiency of our lives, playing a key role in addressing various urban issues. It is predicted to have a worldwide economic impact of 15.7 trillion dollars by 2030 [1]. As a result of the AI revolution, China and the United States will make up about 70% of the world's change. AI technology enhances individuals' understanding and provide them with a deeper sense of the world.

In healthcare, AI is being increasingly utilized, and it is anticipated to have a significant impact on the healthcare industry in the next five or ten years. Professionals in the healthcare field increasingly relying on AI for faster and more accurate diagnoses. Before a patient is admitted to the hospital, AI can assist doctors and nurses in identifying any deterioration in their health. AI is also applied in video games, where AI robots can play complex games like chess, requiring strategic thinking and decision making. Moreover, AI has the capability to analyze data patterns and enhance processes, thereby contributing to the efficiency and sustainability. AI is utilized to detect and prevent potential disasters, such as fires or floods, helping to ensure the safety of citizens and infrastructure. Additionally, AI is applied to analyze and reduce traffic congestion, allowing for better resource management and improved air quality. It is utilized to provide citizens with more tailored services and information, allowing for better public engagement and a more informed citizenry. To improve access to healthcare, providing citizens with more accurate and timely diagnoses and treatments, AI works effectively. AI facilitates the implementation of renewable energy sources, helping cities in reducing their carbon footprint and promoting sustainability.

Businesses in the digital age face the need to safegaurd their data from cyber attacks. One potential solution for ensuring data safety is the utilization of AI. It can assist in detecting and preventing cyber threats, enhancing the overall protection of sensitive information. As an example, the AEG bot and AI2 Platform are used to detect software bugs and cyberattacks with more speed. Facebook, Twitter, and Snapchat have billions of users, and each one needs to be managed and kept safe to ensure it functions properly. A lot of data is organized and managed very well with the assistance of AI. It offers valuable insights into trends, hashtags, and consumer needs by analyzing a lot of data [2]. There is an increase in the use of AI in the travel and transportation industry. All sorts of tasks are performed by AI, from making reservations to suggesting the most efficient routes. It allows the travel industry to provide faster and better service to its customers by acting like a human being. Virtual assistants are now being developed by various car companies to help their customers accomplish their tasks more efficiently. These recent applications of AI play vital role in smart cities as shown in Fig. 6.1.

For example, Tesla Bot, an intelligent virtual assistant, has been created. There are numerous companies currently developing autonomous cars to enhance the security and safety of trips. In the future, intelligent robots may be able to make decisions based on their own experiences rather than being programmed. Several intelligent humanoid robots, such as Erica and Sophia, acts and talk like humans. The use of AI in agriculture is increasing as the industry becomes more digital. The use of robotics in agriculture, sensors to monitor crops and solids, and predictive analyses are all part of this process [3]. Farmers are using AI to help predict crop yields and the best time to plant and harvest crops. It assists farmers to identify pests and diseases and manage irrigation systems more efficiently. AI is also assisting to

Fig. 6.1 Recent utilization of artificial intelligence in sustainable smart cities

automate farm equipment, making it easier and faster to do tasks like plowing, seeding, and harvesting.

E-commerce is becoming more popular due to AI, and it is improving [4]. With AI handling the grading process, academician can allocate more time to teaching. A chatbot that uses AI, assists students in learning by talking to them. AI may one day serve as a virtual teacher for students, allowing them to learn from anywhere and at their own pace. It can also help tailor lessons to individual students, focusing on their weak spots and helping them master the material faster. AI-powered grading offers more accurate grading than human, as AI has the potential to be more consistent and unbiased. This leads to a better learning experience for academicians, teachers, learners and students.

In preliminary data analysis, Fig. 6.2a illustrates the number of documents published per year with the number of publications. The most publications are made in the year 2022 with 160, and as publications are growing each year, this is an excellent field for further research. Articles are the document type with a large number of publications with a percentage of 61.3%. Figure 6.2b shows the document per year by source with subject area. IEEE Access has more publications more than 15. Computer Science subject domain has most of the publications with a percentage of 33%, and Engineering has the second-most publications with a percentage of 24.5%. Figure 6.2c shows the congregation of keywords that are occurring together in a paper. The interrelations between keywords that are co-appearing in the research are linked together. Artificial intelligence, smart city, Internet of things, deep learning, and machine learning are the most co-appearing keywords.

The rest of the chapter is organized as follows: Machine learning/deep learning for sustainable smart cities is presented in Sect. 6.2. The recent applications of artificial intelligence for smart cities are discussed in Sect. 6.3. Existing daily lives example and AI in social media for people is presented in Sects. 6.4 and 6.5, respectively. The open research challenges and research directions are discussed in Sect. 6.6. Finally, Sect. 6.7 concludes with the summary.

6.2 Machine Learning/Deep Learning for Sustainable Smart Cities

Artificial intelligence is a broad term that encompasses a variety of technologies, including machine learning and deep learning. Machine learning is the process of using algorithms to analyze large amounts of data and make predictions. Deep learning is a subset of machine learning that uses artificial neural networks to mimic the workings of the human brain. Relation between artificial intelligence, machine learning, and deep learning is depicted in Fig. 6.3. By using machine learning/deep learning algorithms, cities gathers and analyze data regarding their energy usage, transportation, waste management, and more. These data determines the patterns and make predictions about how to improve the sustainability of the city.

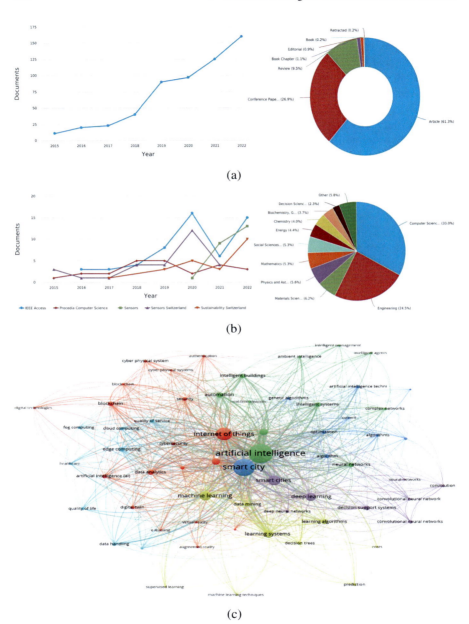

Fig. 6.2 Preliminary data analysis (**a**) documents published per year with the number of publications, (**b**) document per year by source with subject area, (**c**) the congregation of co-appearing keywords from the Scopus database

Fig. 6.3 Relation between artificial intelligence, machine learning, and deep learning

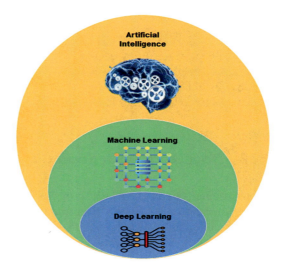

6.2.1 Machine Learning for Smart Cities

Machine learning is a subset of AI that deals with the creation of algorithms that offers learning and make predictions based on data. It help cities to better understand their citizens' needs by collecting data from sources such as sensors, cameras, and social media. Machine learning for smart cities is like having a map of a city. It offers valuable insights into how to navigate the city more efficiently and suggests better routes that may be more scenic or efficient. This data is used to detect patterns, anticipate events, and suggest solutions to improve the quality of life in the city. By leveraging machine learning, cities optimizes their decision-making processes to create efficient urban solutions, creating a more responsive, smarter, and sustainable urban environment. For instance, machine learning algorithms are applied to optimize public transport routes and schedules, predict traffic congestion, and identify areas of pollution. However, some argues that machine learning also have negative consequences. For example, if data from sensors and cameras is used to detect patterns, this could lead to surveillance and a loss of privacy for citizens. Additionally, if machine learning is used to suggest solutions to improve the quality of life in the city, this could lead to a loss of autonomy for citizens. This is because they would be reliant on machines to make decisions for them. Surveillance is the monitoring of the behavior, activities, or other changing information, usually of people for the purpose of influencing, managing, directing, or protecting them.

In some cases, a program's algorithmic procedure differs from a standard process, often referred to as "black box AI" [5]. To train and manage AI systems used by people, many businesses in the European Union are expected to comply with the EU's General Data Protection Regulation [6, 7]. Developing regulations to manage AI will be challenging to establish regulations. In fact, businesses use AI for a variety of reasons since it consists of many different technologies. Limits could potentially impede the development and progress of AI. The following are done with AI-enabled devices:

- To identify a voice, use speech recognition software.
- Already something there needs to be detected.
- Ideas should be developed based on what one have learned.
- AI's ability to learn relies heavily on machine learning.

6.2.2 Deep Learning for Smart Cities

Deep learning is a form of AI that is applied to process large amounts of data. It is utilized to create useful models that is used to make decisions related to smart cities. This technology is capable to monitor traffic, predict crime, optimize energy use, and much more. Deep learning algorithms processes enormous amounts of data rapidly and accurately, allowing cities to make decisions based on the most up-to-date information. It also helps cities to identify patterns in the data that would otherwise be challenging to detect. This can be used to optimize resources and create predictive models to anticipate potential problems before they occur. By leveraging the power of deep learning for smart cities, cities are able to reduce costs, increase efficiency, and improve public safety. As a result, they are very useful in the smart city environment where various sensor data of varying types are collected with the IoT. In addition to feature extraction from sensor data, deep learning algorithms are applied to perform predictions based on feature extraction.

6.3 Recent Applications of Artificial Intelligence for Sustainable Smart Cities

This term "smart cities" refers to the concept of utilizing technologies, including AI, to automate and optimize various aspects of urban life that were traditionally performed by humans. To become a smart city, however, a city must use ICT and AI to achieve long-term social, environmental, and economic growth and improve the quality of life for its people [8–10]. The applications of AI in sustainable smart cities are depicted in Fig. 6.4. Related literature in various applications of smart cities is listed in Table 6.1.

6.3.1 Artificial Intelligence for Smart Transport

The city is already one of the crowded in the world, and AI makes it less congested [11–13]. AI is used to optimize transport systems by reducing congestion, improving safety, and making transportation more efficient. It is also employed to provide personalized and real-time travel information to commuters, helping them better

6.3 Recent Applications of Artificial Intelligence for Sustainable Smart Cities

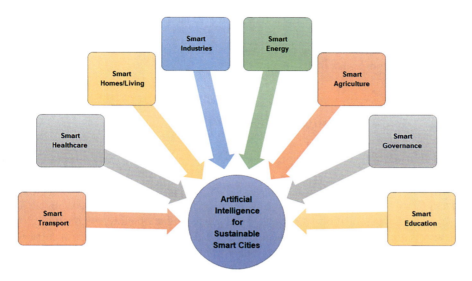

Fig. 6.4 Applications of artificial intelligence in sustainable smart cities

Table 6.1 Reported literatures on artificial intelligence for smart cities components

Component in smart city	Description	References
Smart transport	Smart parking (parking occupancy detection/routing/location prediction)	[14–19]
	Transport management (traffic flow)	[20–22]
	Transport management (traffic accident detection)	[23]
Smart industry	Fault and anomaly detection	[24–36]
	Production management	[37–44]
	Predictive maintenance	[45–48]
Smart homes	Ambient assisted living (activity recognition/fall detection)	[49]
	Ambient assisted living (localization and occupancydetection)	[50, 51]
	energy management (automation, power consumption profiling)	[52–55]
Smart health	Human activity recognition/fall detection	[56]
	Patient health monitoring	[57–61]
	Disease diagnosis	[62–67]
Smart energy	Electricity theft detection	[68]

plan and manage their trips. AI is utilized to offer cost-effective and timely journey options for passengers, enabling them to enjoy a more economical and convenient transportation experience. In addition to parking and traffic management for these vehicles, AI could be helpful in this area as well [14–19]. The use of AI-based traffic management technologies reduces the inconveniences associated with city travel and parking by predicting traffic patterns and parking availability [20–23]. However, there are also potential risks associated with the use of AI in transport. For example, if AI is used to route vehicles, there is a risk that routes may become congested if too many vehicles follow the same AI recommended route. Additionally, if

AI-powered vehicles become too reliant on real-time information, they may be less able to cope with unexpected events, such as traffic accidents or road closures. AI might be used to analyze traffic patterns and allocate resources appropriately based on the company's infrastructure.

6.3.2 Artificial Intelligence for Smart Industries

AI is used to automate processes in various industries, including manufacturing, transportation, healthcare, and retail. It is applied to make decisions based on data, automate tasks, and identify patterns and anomalies. AI is utilized to create virtual agents that interact with humans, providing customer service and carrying out tasks in industrial settings. It is used to detect faults in products and anomalies in industrial processes. By using AI algorithms, machines makes decision faster and more accurately than humans and identifies patterns that may not be readily apparent to human observer. AI is applied to automate tasks, which assists in increasing efficiency and reduce human errors. It offers customer service, as virtual agents which is trained to interact with customers in natural language and provide helpful information.

Detection, monitoring, and a wide range of analyses have all been performed using both machine learning and deep learning methods [26–31]. Other techniques like fog and edge computing are used to detect anomaly detection where edge devices collaboratively train a deep anomaly detection model [32–36]. Production management [37–43] is another application that has found usage of AI in IoT-based smart industry using cloud-based data driven systems along with machine and deep learning algorithms [44–48].

6.3.3 Artificial Intelligence for Smart Home/Smart Living

AI is utilized to automate routine tasks and take care of various daily activities such as controlling lights, setting alarms, and providing recommendations. It saves time and allow users to focus on other more meaningful tasks. With cities committing to a more sustainable future, AI-enabled products and services will become increasingly critical. AI is used to optimize energy consumption, analyze data to detect anomalies, and make decisions based on user preferences and usage patterns. This helps to create a more efficient and sustainable home environment, while lowering costs and protecting the environment.

A number of sites around the world already use AI-driven waste management systems. AI is able to monitor energy consumption in real-time and identify areas of inefficiency and overconsumption. By analyzing data, AI detects anomalies and make decisions on how to optimize energy usage, like when to turn off lights, adjust the temperature, or reduce energy consumption during peak hours. It is also used to automate waste management processes and make sure that waste is disposed of in a more efficient and sustainable way.

6.3.4 Artificial Intelligence for Smart Healthcare

AI in healthcare is used to provide more accurate diagnoses, better personalized treatments and improved drug development. It is used to monitor patient health data, enabling early detection of symptoms and reducing the risk of medical errors. It is utilized to improve patient outcomes and quality of life, with greater efficiency and cost savings.

In smart health, IoT-based AI systems are used for activity recognition/fall detection and disease diagnosis/health monitoring. The goal of activity recognition is to help provide the user with feedback on whether they are getting enough physical exercise or not by using movement sensors such as accelerometers, gyroscopes, and magnetometers. Sports therapy, fall detection, and monitoring illnesses such as Parkinson's and other motor degenerative diseases are among the uses.

For activity detection, inertial sensors are commonly used with various deep learning algorithms in a cloud-based setting. It has been demonstrated that fall detection are performed in fog and edge environments [56]. The authors in [57–67] develop cloud-based health monitoring systems for detecting various types of diseases, such as heart rate, seizure, liver, blood sugar, blood pressure irregular sound, etc. While cloud-based systems offer numerous advantages, they are not exempt from certain limitations. One such drawback is that they require a constant internet connection, which may not be available in all areas. Another drawback is that it is more expensive to maintain the system that are solely based on local hardware.

6.3.5 Artificial Intelligence for Smart Energy

AI is the process of programming a computer to make decisions for itself. This can be done through a number of methods, including but not limited to rule-based systems, decision trees, genetic algorithms, artificial neural networks, and fuzzy logic systems. AI can be utilized to optimize energy consumption in buildings, homes, and factories by analyzing data obtained from sensors and other sources. It has the capability to predict peak energy demand and adjust energy usage accordingly. Additionally, AI can identify and rectify inefficiencies and malfunctions. Moreover, AI can automate energy management processes and provide real-time feedback, empowering users to gain better insights and control over their energy usage. Although AI can be used for all of these things, there is also the potential for it to be exploited maliciously. There is also the potential for AI to be used to hack into energy systems and cause blackouts or other disruptions.

By using AI technologies, cities can use power more efficiently, save money, and improve the quality of life for their residents. It may be possible to control energy distribution in busy places by monitoring and collecting data on people's movements using AI technologies. Using that information, the AI would make energy supply decisions based on energy use patterns throughout the city. A smart grid fault can be classified using decision trees in a cloud-computing system. In [68], the authors identify electricity theft in China based on data from consumers. To detect electricity theft, they employ wide and deep convolutional neural networks to capture periodic and non-periodic components of electricity consumption data.

6.3.6 Artificial Intelligence for Smart Agriculture

In agriculture, AI is used to identify soil problems or nutrient deficiencies. It is used to monitor and analyze environmental factors like temperature, humidity, and soil quality, to help farmers optimize their yields, reduce costs, and minimize potential risks. In order to find and remove weeds, AI may employ robotics, computer vision, and machine learning. The use of AI bots to harvest crops is faster and more efficient than the use of humans. Crop monitoring and disease detection, as well as data-driven crop care and decision-making, are the major applications of AI in IoT for agriculture. AI is utilized to assist farmers, identifing pests and diseases and to help with the decision making in crop management and animal husbandry. Furthermore, AI is applied to predict future market trends, helping farmers to make informed decisions regarding crop selection and pricing. For instance, AI-driven insights assists farmers to make decisions about when to plant and harvest, which crops to plant, and how to price their produce for optimal yields. However, there are some risks associated with using AI in agriculture. For example, if a farmer relies too heavily on AI-driven insights, they may become too reliant on technology and less capable of determining decisions on their own. There is also the potential for data breaches and cyberattacks, which could jeopardize the security of a farm's information.

6.3.7 Artificial Intelligence for Smart Governance

AI is the ability of a computer or machine to perform tasks that typically require human intelligence. This includes visual perception, natural language understanding, and decision-making. It automates with various essential tasks associated with running a government, such as data collection, analysis, and decision-making. It also assists in identifying trends, predict future outcomes, and provide valuable insights to improve public services. By leveraging AI, governments create a smarter policies and offer citizens with a more efficient and streamlined experience. However, there are also concerns that AI will lead to more government surveillance and control. Additionally, there is the potential for AI to be used to manipulate and interfere with democratic processes. As such, it is imperative to ensure that AI is applied in a responsible and transparent manner in order to mitigate these risks. In order to improve governance, it would be helpful to review historical and real-time data on a regular basis. It creates better recommendation engines to connect with customers. The recommendations depend on browsing history and preferences.

6.3.8 Artificial Intelligence for Smart Education

Despite humans' dominance in education, AI is slowly but surely entering the business world. AI is being used to automate administrative tasks, such as grading tests, which frees up teachers' time to focus on their students. AI is also being used to personalize instruction, giving teachers more insight into how each student learns effectively and how to adjust their instruction accordingly. However, AI technology rapidly process large amounts of data and make decisions based on that data, it complete the tasks efficiently and accurately. It also automates repetitive and time-consuming tasks, freeing up educators to focus on more critical tasks. It is utilized to digitize the lecture content, enabling educators to create interactive sessions with students through video conferencing tools. AI is capable to create alternate user interfaces that tailor different grade levels, customed animations and educational content.

AI lesson plans offer students with personalized learning experiences that are tailored to their needs and interests. By using audio and video summaries, students quickly reviews and understand key concepts, and voice assistants provides them with instant answers to their questions. By tracking the student's data, AI algorithms detects patterns in their learning style, allowing for the creation of resources that are tailored to that student's needs. This helps to reduce the cost of creating custom handbooks for every student, while still providing them with the resources they need to succeed.

6.4 Existing Daily Lives Examples of Artificial Intelligence in Smart Cities

Smart cities have the potential to be transformed by AI when utilized wisely. AI provides valuable insights into urban dynamics, enabling cities to become smarter and more efficient. AI plays crucial role in planning and managing transportation networks, optimize energy consumption and enhancing waste management practices. It also has the potential to enhance public safety by predicting crime and issuing warnings when necessary. Additionally, AI actively contributes to enhancing energy efficiency and reducing pollution by facilitating activities such as air quality monitoring, detect problems, and prompt emergency responses. These applications of AI enable cities to access real-time data and analytics, empowering them to make better decisions and ultimately improve the quality of life for citizens. It serves as a vital tool in supporting cities and urban development, and its impact on transforming human lives in recent decades is undeniable. Some real-life examples of AI in smart cities are discussed further.

6.4.1 Cameras

AI-enabled sensors and cameras actively monitor the area around a city, ensuring the safety of its citizens. By using AI-enabled cameras that detect suspicious activities and identify criminals, law enforcement can quickly respond to potential threats. Furthermore, the sensors detect anomalies in the environment, alerting authorities to potential dangers. AI-based security cameras are equipped with sophisticated facial recognition algorithms that detect suspicious behavior in real time. It also identifies objects and tracks their movements, allowing them to identify potential threats before they happen. Additionally, it determines population density, which helps law enforcement better allocate resources to areas where crime is more likely to occur.

6.4.2 Traffic and Parking System

Many commercial vehicles are used in cities to transport people and goods, most of them owned by citizens. These vehicles provide a vital form of transportation for people and businesses and are often the only viable way for people to access goods and services in their area. They also provide an economic boost to the local economy by creating jobs for drivers and businesses. Due to this, AI is useful for tasks such as parking and traffic management.

Real-time maps of parking and traffic are created using road surface sensors or CCTV cameras installed in parking spaces. This would allow drivers to find available spaces quickly, saving them time. It also reduces congestion and pollution while increasing the efficiency of public transportation. Additionally, the data collected from the sensors is used to create a predictive model of traffic flows, allowing for better traffic management and routing. As a result, cars can find available parking spaces more quickly or drive through traffic without a hitch. AI is used to predict where people are traveling, when they will arrive, and what routes they should take to get there. This helps to reduce traffic and congestion, as well as making the overall journey more efficient and faster. Additionally, AI is applied to optimize the pick-up and drop-off points for ridesharing services and to suggest alternate routes for users to reach their destinations in the most efficient way. Cab services like Ola and Uber already use AI to make the trips of their customers easier.

6.4.3 Drones

In order to monitor urban areas, drones that utilize AI technology are deployed. These drones are programmed to recognize certain signs of criminal activity or areas that are prone to flooding or other natural disasters. By using AI, they gather data and alert authorities to any potential problems before they occur. Drones with cameras that provide live images of remote places are helpful for administrative purposes and security. Aerial drones are capable of providing precise maps of areas, tracking traffic, and monitoring the community. Criminal organizations and law enforcement are likely to benefit from the system in terms of security and surveillance.

6.4.4 Face Recognition

Face recognition is achieved through AI, revealing people's unique identities. By using facial recognition technology, security personnel can quickly and accurately identify individuals who may pose a potential threat. This helps deter criminal activity and enable swift apprehension of suspects in the event of a crime. Surveillance cameras or drones equipped with AI that recognize faces and compare them with a database of people to verify identities [69]. Face detection technology of this level is a necessity for smart cities since it gives people more privacy and security. This technology is applied to identify people in public places, improve crime prevention, and reduce the risk of terrorism. It is also used to verify identities in various applications, including access control and border control.

6.4.5 Waste Management

The city waste management problem is also complex due to the number of people living in cities and the amount of garbage they generate. Cities generate more waste than rural areas due to the sheer number of people living in close proximity. This creates large amounts of garbage that must be managed properly in order to keep cities clean and safe. Many countries today are trying to figure out how to deal with the trash that civilization creates while keeping the environment clean and safe. AI-based cameras are capable of identifying and classifying different types of street waste [70]. People could dispose of their trash more easily if AI-enabled sensors were installed on garbage bins. In the case of trash cans, authorities could be alerted when they are nearing capacity, prompting them to adjust routes and schedules garbage management to enhance.

6.4.6 Autonomous Vehicles

It is becoming more common for cars to drive themselves. Autonomous vehicles use a combination of sensors and advanced technology to detect their environment and navigate without a driver. This technology has been shown to greatly reduce the number of traffic accidents, which is why it is becoming increasingly popular. It helps car manufacturers such as Toyota, Audi, Volvo, and Tesla to develop cars that can drive in any environment and avoid accidents. Companies like these use machine learning algorithms to detect obstacles and plan the best trajectory for the car to follow. These algorithms are trained on large datasets of real-world driving data, enabling the car to learn how to recognize and respond to various driving conditions.

6.4.7 Spam Detection Software

Spam detection software utilizes machine learning algorithms to identify spam emails and distinguish them from legitimate emails. It typically does this by analyzing the content of emails and searching for specific words or phrases that are commonly found in spam messages. It also examines patterns in the sender's address and other characteristics that indicate whether an email is spam. The effectiveness of Gmail is email filtering, which successfully filters out 99.9% of spam, is a significant reason for its widespread usage. The machine learning algorithms employed to identify spam emails that rapidly analyze vast quantities of data and identify patterns that may elude human observation. Consequently, the software accurately distinguishes between genuine and spam emails, achieving a remarkably high level of accuracy.

6.4.8 Recognition and Detection of Facial Expressions

This skill is invaluable as it enables us to understand the emotions of others and respond in a suitable manner. By studying facial expressions, we better understand the emotions of those around us and interact with them in a more meaningful way. Face-recognition algorithms are used in smartphones, laptops, and PCs to ensure that only authorized individuals can access them. Citizen's live are becoming more and more influenced by AI every day with the use of face ID and virtual filters in photography. In government buildings and airports, facial recognition plays a role in maintaining safety. These technologies allow us to accurately identify emotions and feelings, even from a distance. Real-time measurement of facial expressions allows for prompt and appropriated response. Moreover, facial recognition technology helps to ensure security and safety in public places, as it is typically resistant to fraudulent attempts.

6.4.9 Navigation

GPS technology helps citizens' stay safe by providing them with complete, accurate, and timely information that is used for navigation. This technology offers users directions to their destination, helps identify the most convenient routes, and alerts them to potential dangers in their vicinity. This helps users stay safe during travel. GPS technology also provides real-time traffic information and get alerts about road closures or accidents that could affect their journey. Moreover, this technology assists users in finding nearby restaurants, gas stations, and other amenities, enhances the convenience of their journey.

6.4.10 Games

In the game industry, AI has gained increasing popularity. By using AI, non-playable characters (NPCs) is designed to act like humans and interact with players. This enhances the realism and engagement of gaming experiences, as NPCs respond to the players' actions in a more natural manner. AI also enables developers to create more complex and unpredictable gaming environments. Software predicts how individuals will react to games during development and testing. Furthermore, AI empowers game developers to design games with more dynamic and adaptive storylines. It facilitates the generation of content on the fly, allowing the game to respond to player choices and provide an ever-evolving gaming experience. This makes the game more engaging and immersive for players, as they are constantly confronted with new challenges, puzzles, and storylines.

6.4.11 Automated Text Corrections and Text Editors

To find errors in word processors, messaging apps, and other text-based media, AI techniques include machine learning, deep learning, and natural language processing. Machine learning algorithms are used for spell-checking and grammar-checking, deep learning algorithms are employed for understanding the context of the text, and natural language processing helps to identify errors in the syntax and semantics of the text. The work is being done by computer scientists and linguists as well. In order to avoid the wrong use comma, editors employ algorithms that have been trained on high-quality linguistic data.

6.5 Artificial Intelligence in Social Media for People in Smart Cities

6.5.1 Instagram

On Instagram, click the explore tab to browse posts that are similar to the interests of users and their followers' interests. The explore tab is powered by an algorithm that takes into account the interests of the users' followers, the posts they have liked, and the posts associated with the hashtags they have used. It then recommends content tailored to user's interests. Based on users' choices and the account they follow, Instagram display posts that align with their interest. The algorithm considers various factors to determine what content is best suited for users. It looks at users' past interactions, the people they follow, the hashtags they use, and other related content to find posts that are similar to their interests or the interests of their followers. This allows Instagram to give users a personalized experience when they use the explore tab.

6.5.2 Facebook

With the help of AI, Facebook is able to better understand what users are talking about. By using natural language processing techniques, Facebook is able to analyze the content of posts and comments, detect the topics and sentiment of conversations, and provide better recommendations based on this data. Automatically translating messages between languages is possible with it. This allows the company to better understand user behavior, provide more relevant ads, and improve its customer service. Additionally, it is used to detect hate speech and other malicious activity, as well as detect trends in conversations, which is utilized to create better user experiences.

6.5.3 Twitter

AI is used to detect patterns in user behavior and content that may be indicative of a scam, false information, or offensive content. This allows Twitter to quickly identify and remove content that violates their terms of service and keep users safe. Additionally, Twitter uses AI to suggest tweets based on what users do with the tweets they see.

6.5.4 Marketing

AI tools such as behavioral analysis and pattern recognition helps marketers target and personalize their ads in a more efficient manner. These tools rapidly analyze vast amounts of data and user behavior, identify patterns, and create highly personalized ads that are tailored to each individual user. This allows marketers to target their ads more effectively, resulting in higher engagement and conversions.

It offers numerous ways to improve content marketing. It enables tracking performance, analysis of campaigns, and various other functionalities. This type of AI analyzes user input and responds as if it were a human, utilizing natural word processing, natural language generation, and natural language understanding techniques. Also, AI enhances marketing strategies by analyzing the actions of clients in real-time.

6.5.5 Chatbots

AI is used to identify and assist customers using the "live chat" option. AI technology is utilized to collect data about the customer's inquiry, analyze it, and provide the customer with the best possible solution. It uses natural language processing to answer customer questions automatically, allowing businesses to provide better customer service with fewer resources. Websites and apps use AI chatbots, which utilize machine learning. AI chatbots creates databases of comments and information. Information is gathered from integrated replies, providing marketers with additional data and insights. As AI becomes more advanced, these chatbots will be able to answer customers' questions and help them around the clock. A customer may be able to benefit from these AI chatbots.

6.5.6 Banking Apps

According to recent surveys, most banks are aware of the benefits of AI. The cutting-edge technology of AI could improve everything from personal finance to commercial finance to consumer financing. Customers who need assistance with wealth management solutions might use SMS text messages or online chats that use AI technology. Unlike humans, AI is capable of detecting changes in transaction patterns and other indications of fraud, resulting in substantial cost saving for businesses and individuals. Additionally, AI provides accurate predictions and assessments of loan eligibilty, detect fraud and automates tasks.

6.6 Open Research Challenges and Research Directions

Artificial intelligence has already changed how people live and work, as well as how they communicate. While these technologies offer numerous benefits to citizens' around the world, they also pose major challenges in the city. The future will be shaped by collaboration between humans and machines. By using the intel assistant, humans are able to make more informed decisions and enhance their own learning. Security assistants plays a crucial role in monitoring security threats, determining relevant information, and devising innovative and dynamic solutions to security issues. Cooperative learning assistants contribute to the continuous learning of students and teachers. In disaster search and recovery operations, human-machine teams collaborate to find survivors, facilitate post-disaster cleanup, and deliver medical care.

AI aids humans in analyzing large amounts of data, detect patterns, and drawing conclusions with greater speed and accuracy compared to humans. This enables us to make better decisions faster and to process and store more data than ever before. It assist humans to automate processes, freeing up time for more creative tasks. Finally, AI facilitates enhanced human connections by providing valuable insights that were previously inaccessible.

- *Artificial Intelligence of Things (AIoT)*

In the blockchain-based robust authentication framework for AIoT, a number of different authentication processes need to be carried out. These authentication processes include verifying the identity of the user, verifying the authenticity of the data, and verifying the integrity of the transaction. All of these processes must be tightly secured in order to ensure the safety and security of the AIoT system. Authentication techniques include 2-factor, 3-factor, and non-certificate methods. As part of the blockchain-designed secure authentication architecture for AIoT, there are many essential components, such as blockchain and AI. They often use the analytical method and deep learning algorithms. It is necessary to have computing power, network bandwidth, and storage space to implement these methods. Meanwhile, these solutions guarantee everyone's safety. Therefore, security systems must be able to withstand a wide range of attacks and intrusions.

- *Interoperability of Tools and Technologies*

These technologies and tools integrate AI, the IoT, and blockchain. In order to achieve this, consensus algorithms, deep learning algorithms, and IoT communication algorithms are all used. The consensus algorithms are used to securely share and store data, while deep learning algorithms are used to identify patterns and make decisions, and IoT communication algorithms are used to connect devices and enable data transfer. There may be a lack of compatibility between tools, technologies, and devices in this environment. IoT devices connected to the internet could be severely affected by this problem. Due to this, it must be handled with care, so more research is needed.

- *Maintaining Your Privacy*

  Securely transmitting data over the internet will prevent hackers from taking advantage of it. In order to protect users' privacy, it is necessary. Private blockchains use encryption and other security measures to restrict access to only those who have permission. This helps protect your data from hackers and prevents it from being accessed without users' consent. Additionally, secure data transmission protocols, such as TLS/SSL, ensure that data is encrypted as it travels across the internet, preventing unauthorized access. In order to maintain users' data safe from individual who do not want them to access it, IoT users should use the most advanced encryption methods available. At the same time, communication entities (such as IoT devices, cloud servers, and users) need to be verified by mutual authentication systems. A security system is employed to prevent unauthorized individuals from accessing restricted areas. Thus, the coder should posses the ability to read and understand all code written. Further, it will be necessary to conduct more research to improve the framework's privacy.

- *Enhancing the Accuracy of the System*

  AI is a major component of these systems, and therefore accuracy values may not be precise. This is because AI-based systems require large amounts of data to train the algorithms, and if the data is inaccurate or outdated, it negatively impact the systems' accuracy. Additionally, AI-based systems are constantly learning, so they may not always make the most accurate decisions. Machine learning and AI users must pay attention to their input and how accurate their information is. Inaccurate algorithms will make it difficult to make accurate predictions. These systems require specific training methods and it is crucial to address these issues to improve their performance.

6.7 Summary

Artificial intelligence's applications are changing businesses and simplifying the resolution of complex problems. For instance, AI contributes to creating more liveable cities by improving security systems, traffic monitoring, and trash management, enhancing people's daily lives, and improving their quality of life. This improvement leads to a higher quality of life and increased safety within cities. AI also helps to optimize energy consumption and reduce emissions, which in turn helps to reduce the impact of climate change. It is employed to improve public services, including healthcare, education, and transportation, allowing cities to operate more efficiently and effectively. Drones, AI security cameras, and face recognition systems need a lot of data from smart cities to train. It is necessary to collect this training data in order to make cities smart enough to last for a long time. AI and machine learning systems based on Cogito's training datasets could be used to build smart cities. It uses computer vision technology to find interesting things in photos and label them as part of its data labeling projects. Gathering this data allows the AI and machine learning systems to recognize patterns, trends, and anomalies in the photos that would otherwise be overlooked. This helps to build the foundations of a smart city, that help to optimize infrastructure, reduce energy consumption, and improve transportation. Connecting with people could be performed through the social networking sites. A number of social networks, including Facebook, Twitter, and Instagram, utilize AI to track users' activity, suggest connections, and show ads to people who are most likely to be interested. The AI system detects and removes content that does not follow the rules using phrases and images. Advertising and marketing professionals utilizing social media sites rely on AI to build their profiles and reach their target audience. AI algorithms recommend similar content to user based on their preferences and interest. Over the last few years, banks have been using AI to make it easier for their customers to pay their bills. Users conveniently deposit money, transfer funds, or even open an account from any location. AI has made it easier for companies to manage their identities, secure their data, and safeguard their privacy. People can even be identified as fraud suspects by watching how they use their credit cards. Using algorithms, we can determine what a person usually buys, when and where they usually buy it, and how much it usually costs.

References

1. PWC, in US$320 billion by 2030? (2022). https://www.pwc.com/m1/en/publications/potential-impact-artificial-intelligence-middle-east.html. Accessed on 8 Mar 2023
2. M. Batty, Artificial intelligence and smart cities. Environ. Plan. B: Urban Anal. City Sci. **45**(1), 3–6 (2018)
3. A.I. Voda, L.D. Radu, Artificial intelligence and the future of smart cities. BRAIN: Broad Res. Artif. Intell. Neurosci. **9**(2), 110–127 (2018)
4. T. Alam, Blockchain cities: The futuristic cities driven by Blockchain, big data and internet of things. GeoJournal **87**(6), 5383–5412 (2021)
5. Z. Allam, Z.A. Dhunny, On big data, artificial intelligence and smart cities. Cities **89**, 80–91 (2019)
6. S. Chatterjee, A.K. Kar, M.P. Gupta, Success of IoT in smart cities of India: An empirical analysis. Gov. Inf. Q. **35**(3), 349–361 (2018)
7. N. Thakur, P. Nagrath, R. Jain, D. Saini, N. Sharma, D.J. Hemanth, Artificial intelligence techniques in smart cities surveillance using UAVs: A survey, in *Machine Intelligence and Data Analytics for Sustainable Future Smart Cities*, Studies in Computational Intelligence, vol. 971, (Springer, 2021), pp. 329–353
8. T. Alam, Blockchain-based big data integrity service framework for IoT devices data processing in smart cities. Mindanao J. Sci. Technol. **19**(1), 1–2 (2021)
9. T. Alam, Cloud-based IoT applications and their roles in smart cities. Smart Cities **4**(3), 1196–1219 (2021)
10. M. Al-Emran, R. Al-Maroof, M.A. Al-Sharafi, I. Arpaci, What impacts learning with wearables? An integrated theoretical model. Interact. Learn. Environ. **30**(10), 1897–1917 (2022)
11. M. Al-Emran, G.A. Abbasi, V. Mezhuyev, Evaluating the impact of knowledge management factors on M-learning adoption: A deep learning-based hybrid SEM-ANN approach, in *Recent Advances in Technology Acceptance Models and Theories*, vol. 335, (Springer, 2021), pp. 159–172

References

12. M.A. Al-Sharafi, N. Al-Qaysi, N.A. Iahad, M. Al-Emran, Evaluating the sustainable use of mobile payment contactless technologies within and beyond the COVID-19 pandemic using a hybrid SEM-ANN approach. Int. J. Bank Mark. **40**(5), 1071–1095 (2022)
13. D. Hema, Smart healthcare IoT applications using AI, in *In Integrating AI in IoT Analytics on the Cloud for Healthcare Applications*, ed. by D. Jeya Mala, (IGI Global, 2022), pp. 238–257
14. X. Sevillano, E. Marmol, V. Fernandez-Arguedas, Towards smart traffic management systems: Vacant on-street parking spot detection based on video analytics, in *Proceedings of the 17th International Conference on Information Fusion (FUSION 2014)*, (Salamanca, Spain, 7–10 July 2014), pp. 1–8
15. D.H. Stolfi, E. Alba, X. Yao, Predicting car park occupancy rates in smart cities, in *Proceedings of the International Conference on Smart Cities*, (Malaga, Spain, 14–16 June 2017), pp. 107–117
16. G. Ali, T. Ali, M. Irfan, U. Draz, M. Sohail, A. Glowacz, M. Sulowicz, R. Mielnik, Z.B. Faheem, C. Martis, IoT based smart parking system using deep long short memory network. Electronics **9**(10), 1696/1–1696/17 (2020)
17. T. Ebuchi, H. Yamamoto, Vehicle/pedestrian localization system using multiple radio beacons and machine learning for smart parking, in *Proceedings of the International Conference on Artificial Intelligence in Information and Communication (ICAIIC)*, (Okinawa, Japan, 11–13 February 2019), pp. 86–91
18. F.M. Awan, Y. Saleem, R. Minerva, N. Crespi, A comparative analysis of machine/deep learning models for parking space availability prediction. Sensors **20**(1), 322/1–322/17 (2020)
19. H. Bura, N. Lin, N. Kumar, S. Malekar, S. Nagaraj, K. Liu, An edge based smart parking solution using camera networks and deep learning, in *Proceedings of the IEEE International Conference on Cognitive Computing (ICCC)*, (San Francisco, CA, USA, 2–7 July 2018), pp. 17–24
20. S.S. Aung, T.T. Naing, Naive bayes classifier based traffic prediction system on cloud infrastructure, in *Proceedings of the International Conference on Intelligent Systems, Modelling and Simulation (ISMS)*, (Kuala Lumpur, Malaysia, 9–12 February 2015), pp. 193–198
21. Y. Xiao, Y. Yin, Hybrid LSTM neural network for short-term traffic flow prediction. Information **10**(3), 105/1–105/22 (2019)
22. W. Wei, H. Wu, H. Ma, An autoencoder and LSTM-based traffic flow prediction method. Sensors **19**(13), 2946/1–2946/16 (2019)
23. N. Dogru, A. Subasi, Traffic accident detection using random forest classifier, in *Proceedings of the 15th Learning and Technology Conference (L and T 2018)*, (Jeddah, Saudi Arabia, 25–26 February 2018), pp. 40–45
24. A.N. Sokolov, I.A. Pyatnitsky, S.K. Alabugin, Research of classical machine learning methods and deep learning models effectiveness in detecting anomalies of industrial control system, in *Proceedings of the Global Smart Industry Conference (GloSIC 2018)*, (Chelyabinsk, Russia, 13–15 November 2018), pp. 1–6
25. J. Windau, L. Itti, Inertial machine monitoring system for automated failure detection, in *Proceedings of the IEEE International Conference on Robotics and Automation*, (Brisbane, Australia, 21–25 May 2018), pp. 93–98
26. M.A. Ferreira, L.F.F. De Souza, F.H.S. Dos Silva, E.F. Ohata, J.S. Almeida, P.P. Filho, Intelligent industrial IoT system for detection of short-circuit failure in windings of wind turbines, in *Proceedings of the International Joint Conference on Neural Networks*, (Glasgow, UK, July 2020), pp. 19–24
27. P.H.F. Sousa, N.M.M. Nascimento, J.S. Almeida, P.P. Reboucas Filho, V.H.C. Albuquerque, Intelligent incipient fault detection in wind turbines based on industrial IoT environment. J. Artif. Intell. Syst. **1**(1), 1–19 (2019)
28. M. Salhaoui, A. Guerrero-Gonzalez, M. Arioua, F.J. Ortiz, A. El Oualkadi, C.L. Torregrosa, Smart industrial IoT monitoring and control system based on UAV and cloud computing applied to a concrete plant. Sensors **19**(15), 3316/1–3316/27 (2019)

29. T. Nkonyana, Y. Sun, B. Twala, E. Dogo, Performance evaluation of data mining techniques in steel manufacturing industry. Procedia Manuf. **35**, 623–628 (2019)
30. J.E. Siegel, S. Pratt, Y. Sun, S.E. Sarma, Real-time deep neural networks for internet-enabled arc-fault detection. Eng. Appl. Artif. Intell. **74**, 35–42 (2018)
31. D.C. Huang, C.F. Lin, C.Y. Chen, J.R. Sze, The Internet technology for defect detection system with deep learning method in smart factory, in *Proceedings of the 4th International Conference on Information Management (ICIM 2018)*, (Oxford, UK, 25–27 May 2018), pp. 98–102
32. L. Li, K. Ota, M. Dong, Deep learning for smart industry: Efficient manufacture inspection system with fog computing. IEEE Trans. Industr. Inform. **14**(10), 4665–4673 (2018)
33. S.Y. Lin, Y. Du, P.C. Ko, T.J. Wu, P.T. Ho, V. Sivakumar, R. Subbareddy, Fog computing based hybrid deep learning framework in effective inspection system for smart manufacturing. Comput. Commun. **160**, 636–642 (2020)
34. M.S.S. Garmaroodi, F. Farivar, M.S. Haghighi, M.A. Shoorehdeli, A. Jolfaei, Detection of anomalies in industrial IoT systems by data mining: Study of CHRIST Osmotron water purification system. IEEE Internet Things J. **8**(13), 10280–10287 (2020)
35. Y. Liu, N. Kumar, Z. Xiong, W.Y.B. Lim, J. Kang, D. Niyato, Communication-efficient federated learning for anomaly detection in industrial internet of things, in *Proceedings of the GLOBECOM*, (Taipei City, Taiwan, 7–10 December 2020), pp. 1–6
36. D. Park, S. Kim, Y. An, J.Y. Jung, LiReD: A light-weight real-time fault detection system for edge computing using LSTM recurrent neural networks. Sensors **18**(7), 2110/1–2110/15 (2018)
37. H. Zheng, Y. Feng, Y. Gao, J. Tan, A robust predicted performance analysis approach for data-driven product development in the industrial internet of things. Sensors **18**(9), 2871/1–2871/16 (2018)
38. A. Essien, C. Giannetti, A deep learning model for smart manufacturing using convolutional LSTM neural network autoencoders. IEEE Trans. Industr. Inform. **16**(9), 6069–6078 (2020)
39. B. Huang, W. Wang, S. Ren, R.Y. Zhong, J. Jiang, A proactive task dispatching method based on future bottleneck prediction for the smart factory. Int. J. Comput. Integr. Manuf. **32**(3), 278–293 (2019)
40. W. Tao, Z.H. Lai, M.C. Leu, Z. Yin, Worker activity recognition in smart manufacturing using IMU and sEMG signals with convolutional neural networks. Procedia Manuf. **26**, 1159–1166 (2018)
41. A.R.M. Forkan, F. Montori, D. Georgakopoulos, P.P. Jayaraman, A. Yavari, A. Morshed, An industrial IoT solution for evaluating workers' performance via activity recognition, in *Proceedings of the 2019 IEEE 39th International Conference on Distributed Computing Systems (ICDCS)*, (Dallas, TX, USA, 7–10 July 2019), pp. 1393–1403
42. T. Wang, X. Wang, R. Ma, X. Li, X. Hu, F.T. Chan, J. Ruan, Random Forest-Bayesian optimization for product quality prediction with large-scale dimensions in process industrial cyber-physical systems. IEEE Internet Things J. **7**(9), 8641–8653 (2020)
43. Y. Han, C.J. Zhang, L. Wang, Y.C. Zhang, Industrial IoT for intelligent steelmaking with converter mouth flame spectrum information processed by deep learning. IEEE Trans. Industr. Inform. **16**(4), 2640–2650 (2020)
44. B. Brik, B. Bettayeb, M. Sahnoun, F. Duval, Towards predicting system disruption in industry 4.0: Machine learning-based approach. Procedia Comput. Sci. **151**, 667–674 (2019)
45. Y. Qu, X. Ming, S. Qiu, M. Zheng, Z. Hou, An integrative framework for online prognostic and health management using internet of things and convolutional neural network. Sensors **19**(10), 2338/1–2338/14 (2019)
46. S. Hwang, J. Jeong, Y. Kang, SVM-RBM based predictive maintenance scheme for IoT-enabled smart factory, in *Proceedings of the 2018 13th International Conference on Digital Information Management, ICDIM 2018*, (Berlin, Germany, 24–26 September 2018), pp. 162–167
47. E.J.K. Jung-Hyok Kwon, Failure prediction model using iterative feature selection for industrial internet of things. Symmetry **12**(3), 454/1–454/12 (2019)

48. W. Zhang, W. Guo, X. Liu, Y. Liu, J. Zhou, B. Li, Q. Lu, S. Yang, LSTM-based analysis of industrial IoT equipment. IEEE Access **6**, 23551–23560 (2018)
49. G.Y. Kim, S.S. Shin, J.Y. Kim, H.G. Kim, Haptic conversion using detected sound event in home monitoring system for the hard-of-hearing, in *Proceedings of the HAVE 2018—IEEE International Symposium on Haptic, Audio-Visual Environments and Games*, (Dalian, China, 20–21 September 2018), pp. 17–22
50. A.B. Adege, H.P. Lin, G.B. Tarekegn, S.S. Jeng, Applying deep neural network (DNN) for robust indoor localization in multi-building environment. Appl. Sci. **8**(7), 1062/1–1062/14 (2018)
51. L. Zimmermann, R. Weigel, G. Fischer, Fusion of nonintrusive environmental sensors for occupancy detection in smart homes. IEEE Internet Things J. **5**(4), 2343–2352 (2018)
52. F.C.C. Garcia, C.M.C. Creayla, E.Q.B. Macabebe, Development of an intelligent system for smart home energy disaggregation using stacked denoising autoencoders. Procedia Comput. Sci. **105**, 248–255 (2017)
53. I.C. Konstantakopoulos, A.R. Barkan, S. He, T. Veeravalli, H. Liu, C. Spanos, A deep learning and gamification approach to improving human-building interaction and energy efficiency in smart infrastructure. Appl. Energy **237**, 810–821 (2019)
54. D. Popa, F. Pop, C. Serbanescu, A. Castiglione, Deep learning model for home automation and energy reduction in a smart home environment platform. Neural Comput. Applic. **31**, 1317–1337 (2019)
55. C.C. Yang, C.S. Soh, V.V. Yap, A non-intrusive appliance load monitoring for efficient energy consumption based on Naive Bayes classifier. Sustain. Comput. Inf. Syst. **14**, 34–42 (2017)
56. D. Yacchirema, J.S. de Puga, C. Palau, M. Esteve, Fall detection system for elderly people using IoT and ensemble machine learning algorithm. Pers. Ubiquit. Comput. **23**, 801–817 (2019)
57. I. Machorro-Cano, G. Alor-Hernandez, M.A. Paredes-Valverde, U. Ramos-Deonati, J.L. Sanchez-Cervantes, L. Rodriguez-Mazahua, PISIoT: A machine learning and IoT-based smart health platform for overweight and obesity control. Appl. Sci. **9**(15), 3037/1–3037/23 (2019)
58. A. Alamri, Monitoring system for patients using multimedia for smart healthcare. IEEE Access **6**, 23271–23276 (2018)
59. M. Awais, M. Raza, N. Singh, K. Bashir, U. Manzoor, S. ul Islam, J.J. Rodrigues, LSTM based emotion detection using physiological signals: IoT framework for healthcare and distance learning in COVID-19. IEEE Internet Things J. **8**(23), 16863–16871 (2020)
60. M. Alhussein, G. Muhammad, M.S. Hossain, S.U. Amin, Cognitive IoT-cloud integration for smart healthcare: Case study for epileptic seizure detection and monitoring. Mob. Netw. Appl. **23**, 1624–1635 (2018)
61. M.S. Hossain, G. Muhammad, Cloud-assisted industrial internet of things (IIoT)—Enabled framework for health monitoring. Comput. Netw. **101**, 192–202 (2016)
62. M. Ganesan, N. Sivakumar, IoT based heart disease prediction and diagnosis model for healthcare using machine learning models, in *Proceedings of the 2019 IEEE International Conference on System, Computation, Automation and Networking, ICSCAN 2019*, (29–30 March 2019), pp. 1–5
63. S. Mohapatra, P.K. Patra, S. Mohanty, B. Pati, Smart health care system using data mining, in *Proceedings of the 2018 International Conference on Information Technology, ICIT 2018*, (December 2018), pp. 44–49
64. P. Kaur, R. Kumar, M. Kumar, A healthcare monitoring system using random forest and internet of things (IoT). Multimed. Tools Appl. **78**, 19905–19916 (2019)
65. S. Tuli, N. Basumatary, S.S. Gill, M. Kahani, R.C. Arya, G.S. Wander, R. Buyya, HealthFog: An ensemble deep learning based smart healthcare system for automatic diagnosis of heart diseases in integrated IoT and fog computing environments. Futur. Gener. Comput. Syst. **104**, 187–200 (2020)
66. M. Alhussein, Monitoring Parkinson's disease in smart cities. IEEE Access **5**, 19835–19841 (2017)

67. M. Devarajan, L. Ravi, Intelligent cyber-physical system for an efficient detection of Parkinson disease using fog computing. Multimed. Tools Appl. **78**, 32695–32719 (2019)
68. Z. Zheng, Y. Yang, X. Niu, H.N. Dai, Y. Zhou, Wide and deep convolutional neural networks for electricity-theft detection to secure smart grids. IEEE Trans. Ind. Infor. **14**(4), 1606–1615 (2018)
69. J. Aguilar, A. Garces-Jimenez, M.D.R. Moreno, R. Garcia, A systematic literature review on the use of artificial intelligence in energy self-management in smart buildings. Renew. Sust. Energ. Rev. **151**, 111530/1–111530/16 (2021)
70. T.M. Ghazal, M.K. Hasan, M.T. Alshurideh, H.M. Alzoubi, M. Ahmad, S.S. Akbar, I.A. Akour, IoT for smart cities: Machine learning approaches in smart healthcare—A review. Fut. Internet **13**(8), 218/1–218/19 (2021)

Chapter 7
Energy Management of Sustainable Smart Cities Using Internet-of-Energy

7.1 Introduction

Internet of Things (IoT) devices, comprising sensors, actuators, communication, and networking technology, play a crucial role in smart city solutions. These solutions address the challenges of urbanization and population growth, particularly in the context of energy management [1]. The focus on sustainable smart cities necessitates energy efficient solutions that benefit both citizens and the environment. Over the past decade, energy applications has increased due to the rising demand in various sectors such as automation, factories, transport, domestic appliances, healthcare, transport, and other services in smart cities as shown in Fig. 7.1. While progress has been made in expanding access to power, increasing renewable energy usage, and improving energy efficiency globally, there is still a need to deliver affordable, reliable, sustainable, and modern energy for all. Additionally, with the projected urban population reaching 68.4% of the total population by 2050, it is crucial to address not only the escalating consumption of fossil fuels in cities but also the excessive load on local power plants due to energy consumption imbalances [2]. Buildings and infrastructure account for over 40% of worldwide energy usage, with the residential sector contributing approximately 7%. Therefore, technological ideas that successfully save energy in homes are of considerable interest [3]. With an average annual electrification rise of 0.876% points, global access to electricity climbed from 83% in 2010 to 90% in 2019. The worldwide access deficit has decreased from 1.22 billion in 2010 to 759 million in 2019. In 2030, it is possible that 660 million people still have no access to the internet despite the tremendous efforts made [4]. Implementing the 2030 agenda for sustainable development [5] requires the development of comprehensive indicators and statistical data to monitor progress, guide policy, and enhance accountability. On July 6, 2017, the United Nations General Assembly endorsed the global indicator framework, which

© The Author(s), under exclusive license to Springer Nature Switzerland AG 2023
P. Mishra, G. Singh, *Sustainable Smart Cities*,
https://doi.org/10.1007/978-3-031-33354-5_7

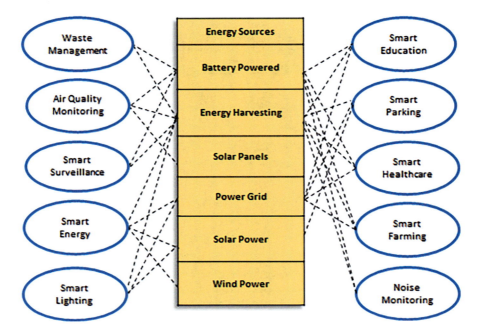

Fig. 7.1 Applications of smart cities in energy sector

was also incorporated into the Assembly's Resolution on the Statistical Commission's Work in support of the 2030 agenda for Sustainable Development [6].

Information and communication technologies have become major energy consumers due to generating billions of requests that take many forms, including emails, audio-video transmissions, search engine results, social media, etc. A stable and sustainable energy source is required for the successful establishment of smart cities [7].

Both fossil fuels and renewable energy are important sources of energy, but due to the significant environmental factors in the development of smart cities, renewable energy sources are the best options for addressing smart city energy needs in the future [8]. In the construction of smart cities, policymakers, and urban planners should prioritize the use of abundant renewable energy sources. It should be emphasized that simply providing numerous energy sources for customers is insufficient, and policymakers and urban planners should consider other issues such as affordability and availability in the context of sustainability [9]. In this regard, determining a smart plan for public transportation systems, such as the deployment of IoT technology, is necessary in order to alleviate traffic congestion in places where the population is booming. On the other hand, since renewable energy sources are critical resources for future global development, particularly in the face of climate change, the integration of new emerging technologies such as artificial intelligence (AI) has the potential to aid in the future resolution of energy sustainability

7.1 Introduction

challenges. However, providing sustainable energy for a smart city, will require an integrated infrastructure that incorporates emerging technologies such as the IoT and next-generation mobile communication.

For energy supply, transmission, distribution, and demand, this plan is more useful [10]. Also, to change the strategies for energy provision to larger settlements, the use of energy systems with new technological options such as photovoltaic (PV)-driven heat pumps for heat provision, bio-methane injection into grids, passive buildings, and small-scale combined heat and power (CHP) with heat storage is appropriate and effective [11].

In addition, strategies and initiatives such as

- Use of electric vehicles (EVs) [12].
- Evaluation of common frameworks for interaction between intelligent transportation and EVs in smart cities [13].
- Regular and reasonable electricity pricing strategy that contributes to grid security [14].
- Adequate investment and better government support [15].

These points are necessary to create intelligent energy systems for smart cities. The management of energy in smart cities has become an increasingly important topic in industry and research. The International Telecommunication Union (ITU) in collaboration with the IEEE, has established a smart cities community to assist governments in making the switch to smart cities. A number of firms, including Honeywell, IBM, Intel, Cisco, and Schneider Electric, are working on energy-efficient smart city solutions. Research funded by the Seventh Framework Program (FP7) of the European Commission has focused on energy-efficient smart cities over the last few years.

Smart devices powered by the internet provide automated and efficient solutions that improve efficiency. Moreover, because massive amounts of these small devices are constantly sharing data, the IoT faces a significant challenges in terms of coverage, safety, complexity, expenses, and energy usage. This calls for an approach to energy preservation for cost-effective and need-base resource usage. In addition, IoT has significantly improved society's energy needs, sustainable development, performance, demands, and security. It has now been extended to the industrial sector through Industry 4.0, for productivity improvements, management, and meeting high energy demands [16]. A cloud-based energy management solution for smart buildings was recently launched [17], in an effort to address energy management issues for industries. A smart city can be made more sustainable and greener through IoTs enabled smart solutions with ALMANAC: Reliable Smart Secure Internet of Things for Smart Cities [18]. The project, "Reliable, Resilient, and Secure IoT for Smart City Applications," aims to develop, evaluate, and test a framework for IoT-enabled smart city applications so that smart devices may behave in a more energy-efficient manner [19]. Some important projects in energy sector are discussed in Table 7.1.

The energy system model has been established for decades and, as noted in [20], is constantly evolving to incorporate new concepts and technology. Specifically, this

Table 7.1 List of projects involved in cities

Sr. no.	Project name	References	Year	City involved	Purpose
1	STEEP (Systems Thinking for Efficient Energy Planning)	[59]	2013–2015	San Sebastian (Spain), Bristol (UK) and Florence (Italy).	The possibilities of employing PVs, cogeneration setups, geothermal and hydrothermal heat pumps, short- and long-term thermal energy-storage systems, waste heat recovery from industry, and smart meters were investigated
2	STEP UP (Strategies Towards Energy Performance and Urban Planning)	[60]	2012–2015	Ghent (Belgium), Glasgow (Scotland), Gothenburg (Sweden) and Riga (Latvia)	Employing CHP, renewables in terms of biomass, geothermal, and solar thermal was a major section of the project
3	PLEEC (Planning for Energy- Efficient Cities)	[61]	2014–2016	(Sweden), Turku (Finland), Santiago de Compostela (Spain), Jyväskylä (Finland), Tartu (Estonia) and Stoke-on-Trent (Estonia)	The used technologies were water management, electrical power grids, heating and cooling grids, and renewables
4	ZenN (Nearly Zero Energy) project	[62]	2013–2017	Oslo (Norway), Malmö (Sweden), Eibar (Spain) and Grenoble (France)	Aimed to reduce energy use in existing buildings and neighborhoods. Photovoltaic and solar thermal panels besides ground, air and solar source heat pumps, and cogeneration units had major roles in the project
5	R2CITIES	[63]	2013–2018	Valladolid (Spain), Genoa (Italy) and Istanbul (Turkey)	Photovoltaic energy, solar thermal, storage systems, and distribution systems were among the utilized technologies for achieving the project targets. Cities for achieving nearly zero-energy cities
6	READY	[64]	2014–2019	Aarhus (Denmark) and Växjö (Sweden)	Developed integrated smart city electric grid systems and mobility solutions. They employed integrated PV power and heat generation, batteries for electricity storage, district heating and electricity systems, and intelligent control (by Power Hub) of energy consumption and production for optimizing the utilization of renewable energy resources more efficiently and reducing fuel consumption

7.1 Introduction

type of model is commonly used for the planning, operation, and management of power systems. In [21], the authors proposed a distribution network growth-planning model that took into account the size, location, and timing of DG investments and network enhancements. In [22], a smart grid using software agents is simulated in order to simulate the dynamic behavior of a smart city while considering only electricity. Modeling an electric system is usually done using stochastic programming, which involves minimizing an objective function while considering certain limitations, as stated in [23]. For this reason, urban planning models are necessary to achieve sustainable development [24]. In [25], a model is presented for evaluating the availability of renewable energy and urban energy supply plants at various sites to determine the optimal locations and types of generation to install given geographical constraints. Only a few cities approach urban planning through a holistic strategy that enables synergy in energy-related activities at various scales, rather than prioritizing renewable energy or energy efficiency [26]. Similarly, [27] describes methods for enhancing energy efficiency in urban design using information technology.

Authors in [28] propose a challenge of organizing many groups of cumulative air conditioners for delivery system load management. In [29], an Artificial Neural Network for a home energy management system based on bluetooth low energy was described. Infrastructure technology aims to achieve significant energy savings in a variety of contexts, including infrastructure, development, transportation, buildings, electricity generation, and delivery, in order to reduce greenhouse gas emissions and provide effective ecological security. Similarly, in [30] proposes a smart grid-based human-centric smart home energy management system. An improved evolutionary algorithm is used to plan an Intelligent Residential Energy Management System (IREMS) for housing buildings in order to reduce electrical energy expenses [31]. In [32] a Neural Network (NN)-based Housing Convenient Demand Profile Estimator (HCDPE), which has been developed as a practical Energy Hub Management System (EHMS) was presented. Also, [33] presents two power management solutions for making V2G practicable for plug-in electric vehicles (PEVs) in grid-associated microgrids with a supervisory power imbalance. Authors in [34] describe a smart, energy-efficient system in smart grid Cyber-Physical Systems (CPSs) using coalition-based game theory. In [35], smart-grid power-driven coordinated multipoint (CoMP) transmission with active energy management was suggested. An infinite-horizon optimization problem is formulated to obtain the most advantageous downlink that is robust to control reservations [36]. proposes an Intelligent Dynamic Energy Management System (I-DEMS) for a smart microgrid. A distributed Home Energy Management System with Storage (HoMeS) that comprises compound microgrids and several clients is presented in [37]. In [38], a power system involving distributed generators and consumers, as well as a Unified Energy Management System (UEMS) based on Distributed Location Marginal Pricing (DLMP), was proposed. In this chapter, we acknowledge energy management for Internet of Energy (IoE)-based sustainable smart cities with the following points:

- Energy management classification for IoE-based sustainable smart cities in terms of scheduling optimization, low power device transceivers, cognitive framework, and cloud computing technology.
- Energy harvesting in smart cities focuses on receiver design, energy optimization, energy sources, energy scheduling, and energy routing.
- Sustainable smart cities in various domains like conceptual framework and enabling technologies based on energy management.
- Open research challenges and future research directions for energy management in sustainable smart cities enabled by the Internet of Things.

The current chapter also demonstrated the importance of smart grids as a major component of establishing smart cities by providing an overview of the key constituents of the smart grids in terms of micro/nano-grids. The rest of the chapter is organized as follows: Sects. 7.2 and 7.3 describes recent advancement in energy sector and the classification of energy management for IoE in smart cities, respectively. Energy harvesting in smart cities is presented in Sect. 7.4. Sections 7.5 and 7.6 explains the green energy and conceptual framework, respectively. Enabling technologies in energy sector is discussed in Sect. 7.7. Open research challenges and research directions are described in Sect. 7.8. Eventually, a summary is provided in Sect. 7.9.

7.2 Recent Advances in Energy Sector

7.2.1 Internet of Energy

Internet of Energy is a term coined by the IoT, which refers to interconnected devices, big data analytics, and people-to-people (P2P), machine-to-people (M2P), and machine-to-machine (M2M) communication. Based on the IoTs, the IoE provides information to enable control and optimization of the power grid. The idea is to make the electricity grid more self-sufficient. By using IoT devices such as smart sensors and communication technology, the energy industry is establishing the IoE to manage energy production and resources. The IoE is basically a smart energy infrastructure system that connects every point in the power grid, including generation, load, distribution, storage, and smart meters, with the IoT as depicted in Fig. 7.2. The IoT makes the electrical grid more efficient, reliable, and resilient. Instead of a two-way flow of information, the IoT allows for a multidirectional flow.

Based on Cisco's value at stake calculations, Cisco examines several public sector use cases, including education, culture and entertainment, transportation, safety and justice, energy and environment, healthcare, defense, and next-generation work [39] as shown in Fig. 7.3. As smart gadgets have grown in popularity, the IoE has opened up the possibility of low energy loads with increased internet connectivity, allowing for minimized energy consumption through smart, practical, and controlled use. Using the IoT, real-time applications runs with minimal inputs and

7.2 Recent Advances in Energy Sector

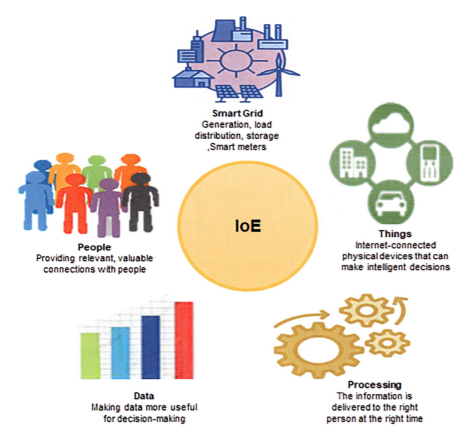

Fig. 7.2 Concept of Internet of Energy

effort in the energy sector, enabling large-scale digital change. The technology is useful for industrial processes that require connectivity, safety, flexibility, accuracy, reliability, automation, and minimal latency. However, the interaction between humans and devices generates large amounts of data, which necessitates significant energy consumption in terms of storage and data modeling. This poses challenges in terms of data exchange, security, and performance.

It refers to the modernization and digitization of energy producers' and manufacturers' electricity systems. With this technology, energy generation becomes more effective and clean, resulting in less waste. Improved efficiency, cost reductions, and lower energy consumption are all advantages of implementing IoE. Manufacturers and producers eliminates inefficiencies in the current energy infrastructure by increasing power generation, transmission, and utilization of IoE technology. By making improvements to electric infrastructures, the flow of energy can reach its full potential, reducing energy waste. However, without important updates and efficient information transfer, a significant amount of energy can be lost along the line.

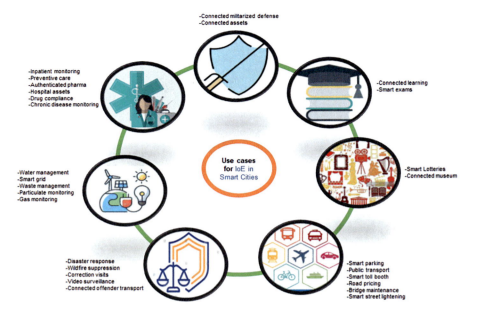

Fig. 7.3 Use cases for IoE in sustainable smart cities

If an IoE system is not implemented, energy may be wasted when transmitted across lines are unable to convey it efficiently. By incorporating IoE technology into the process, smart grid technology is also installed [40].

A sustainable smart city is a network of connected devices that enables residents to experience greater convenience, security, and ultimately, a higher quality of life. Energy is one aspect of the smart city that has rapidly evolved in recent years. Effective energy consumption is essential for urban living. Electric vehicles play a significant role in mitigating climate change by reducing carbon emissions. Smart meters have already become familiar to a large number of consumers worldwide. These devices are designed to directly connect home's electricity or gas meter to the energy distributor.

7.2.2 Energy 4.0

Advanced technologies and the integration of the IoT into industries have given rise to the concept of Energy 4.0. Energy consumption plays a significant role in global emissions and climate change, leading to increased cost. In response, smart grids, renewable energy management, and distributed generation are being developed as part of the digital revolution known as Energy 4.0. Developers of hardware/software are focusing on creating industry-driven solutions that prioritize reliability and environmental safety. Digitalization provides opportunities for Energy 4.0 companies to develop new business models and sustainable energy solutions, with costs decreasing and technology advancing rapidly. Consequently, there is a growing need to address the quantity and nature of energy consumption across all sectors and to explore new innovative solutions.

The advantages of new electricity sources such as wind and solar have been proven to be mature and reliable technology. However, there are still obstacles that need to be overcome. The next step is to optimize these new generation sources to align with consumer consumption patterns. Energy 4.0 represents a significant progression, similar to the Industrial Revolution. The optimization of assets using big data and AI has enabled a more connected and intelligent asset ecosystem. This has the potential to greatly reduce electricity consumption and greenhouse gas emissions. By leveraging new renewable generation and improving the energy efficiency of existing assets and operations makes this possible. While there are security and legal concerns to address, there are already notable examples of Energy 4.0 in action worldwide, demonstrating the benefits of this digitalization movement.

7.2.3 Smart Grid

Smart grids, which are upgraded versions of conventional energy infrastructures, are essential parts of a sustainable smart city design [41] as depicted in Fig. 7.4. By connecting smart grids and buildings for efficient energy consumption and production, the next generation of smart energy systems and grids will be able to control the energy of buildings [42]. In reality, by utilizing available resources, the smart grid provides new services to residents of smart homes [43]. Smart meters enable users to understand and contribute to carbon emissions, as well as assist them in making better decisions [42–44].

The energy industry, on the other hand, will be able to better manage energy supply and demand. Additionally, peak demand management will allow consumers to manage high wattage appliances such as air conditioners, electric water heaters, pool pumps, clothes dryers, etc. With the increased popularity of electric vehicles (EVs), smart grids are able to detect and accept generated or stored energy from customers.

The smart grid is widely regarded as one of the most essential uses of the IoT [45–47]. The smart grid relies heavily on the transmission, perception, and processing of information. Electricity grids are integrated with IoT technologies in a variety of ways, including controlling, collecting, monitoring, and constructing. Smart grids are intended to be connected to a variety of linked devices capable of providing large volumes of data about the network's status [48]. A cloud-based system can handle computing, real-time processing, and optimizing huge volumes of data in a cost-effective manner. The remaining needs are met by utilizing IoT technologies [49, 50], edge resources, and integrating the cloud into a standardized architecture.

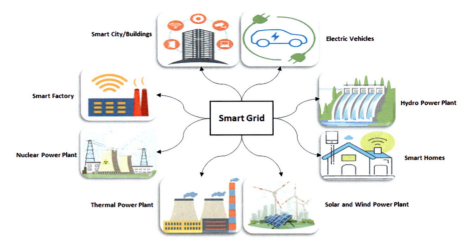

Fig. 7.4 Applications of smart cities in smart grid

Furthermore, status monitoring enables smart grid control, automation, problem identification, management, and forecasting. With the smart grid, two new concepts for the electric system have been introduced: microgrids and nano-grids.

- A microgrid is a system for powering a small group of consumers (residential complexes, hospitals, data centers, schools, etc.) with renewable energy sources such as solar photovoltaics (PV), wind power, microturbines, and fuel cells [51]. A microgrid is a modest self-contained power system that connects local loads with distributed generation with a power capacity of 10–100 kW [52, 53].
- Unlike microgrids, nano-grids have a similar composition of local generation based on renewable resources and energy storage, but their power output is up to 25 kW, and they are designed for residential or commercial buildings.

Some major issues, including energy losses in long distance transportation [54], and their associated costs, are reduced by producing and distributing energy through intelligent energy systems. This is accomplished through smart load and building automation, which improve energy efficiency, safety, and comfort on both home and industrial scales [55–57]. In [58], the focus is on the emerging challenges in urban areas, with particular emphasis on the importance of developing intelligent grids. They said that smart grids would provide benefits as follows:

- Resolving the energy crisis
- Promoting sustainable urban growth
- Reducing energy consumption
- Providing a flexible power market mechanism
- Allowing for extensive client participation

7.3 Classification of Energy Management in Sustainable Smart Cities

In smart cities, the number of IoE applications is increasing, resulting in more energy-efficient devices. Many energy saving technologies available are capable of reducing energy consumption or maximizing resource efficiency. The following are some of the most significant research trends related to energy-efficient IoT-enabled smart cities: optimization of scheduling, implementation of low power transceivers, creation of cognitive frameworks, and deployment of cloud computing technologies [65–67] as depicted in Fig. 7.5.

7.3.1 Scheduling Optimization

A smart city that employs IoE-enabled scheduling optimization aims to reduce energy consumption and resulting electricity consumption by optimizing available resources. It is important to note that demand side management (DSM) involves in the usage of residential electricity to reduce costs by altering the system load shape. DSM consists of two major components: load shifting and energy conservation. Load shifting occurs when a customer's load shifts from a high-peak state to a low-peak state. This approach helps to save electricity and accommodate additional users simultaneously. Energy conservation focuses on reducing overall energy usage by implementing efficient technologies, promoting energy-saving practices, and encouraging behavioral changes. Together, these components play a vital role in achieving energy efficiency and sustainable resource management.

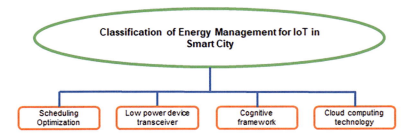

Fig. 7.5 Classification of energy management for IoT in smart cities

7.3.2 Low Power Device Transceivers

Despite the power constraints of IoT devices in smart city applications, a low power conceptual model is necessary to manage energy usage in IoE-based smart cities. Most of the application protocols for IoT devices are not compatible with energy efficiency. The radio duty cycle, in particular, is a key factor for IoT device energy efficiency, and researchers are exploring ways to reduce radio duty cycles and design IoT devices that are more energy efficient.

7.3.3 Cognitive Framework

As IoT devices are heterogeneous, the services associated with them can be unpredictable. Therefore, it is critical that smart cities incorporate intelligence and cognitive techniques throughout the IoT architecture. It is vital that the framework includes reasoning and learning capabilities to improve IoE network decisions. In [68], a cognitive management architecture was described that made decisions based on the contextual background of IoT devices.

7.3.4 Cloud Computing Technology

With the rise of cloud computing and storage, IoE-enabled smart cities provide energy-efficient solutions. More specifically, a cloud-based strategy also improves the efficiency of data centers by managing enormous flexibility.

7.4 Internet of Energy Harvesting Things in IoE-Based Smart Cities

Energy harvesting has been suggested as a way to extend the life of IoE equipment in smart cities. IoT offers an effective means for monitoring and managing the globe from anywhere at any time. This concept requires a large number of sensors to collect data with high accuracy over a large service area. Replacing the batteries of millions of devices is extremely difficult, the Internet of Energy Harvesting Things is being proposed as a solution to this problem. Due to this, IoT devices in smart cities need to overcome several challenges, which are outlined below as shown in Fig. 7.6.

Fig. 7.6 Internet of Energy harvesting Things in IoE-based smart cities

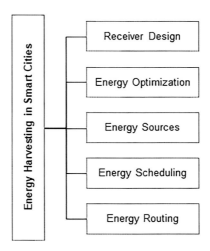

7.4.1 Receiver Design

The fundamental challenge of radio frequency-based energy harvesting is the design of energy harvesting circuits. The harvesting circuit's sensitivity is higher than that of standard receivers, which leads to oscillations in energy transmission due to environmental and mobile factors. Optimal energy harvesting requires a reliable and effective circuit design. Therefore, circuit designers should incorporate advanced technologies to enhance the efficiency of RF to DC conversions.

7.4.2 Energy Optimization

The degree of uncertainty associated with energy generation from ambient sources is higher compared to the degree of uncertainty associated with focused energy generation. This is because ambient sources utilize renewable energy, while focused energy sources are predetermined by network designers based on the harvesting requirements of IoT devices. To examine the performance of energy harvesting systems in smart cities, it is crucial to model the energy arrival rate accurately and comprehensively.

7.4.3 Energy Sources

IoT devices located far from energy sources may experience challenges in harvesting efficiently. Consequently, devices that are distant from dedicated power sources may deplete their energy, leading to a reduced network lifespan. While ambient energy sources cannot be altered, the placement and number of dedicated energy sources must be carefully considered in the case of dedicated energy harvesting.

7.4.4 Energy Scheduling

By implementing assignment-based energy harvesting, where devices are scheduled for RF power transmission according to their harvesting demands, it is possible to reduce energy consumption. However, in order to achieve this, a certain level of coverage and sufficient harvesting time are required. Therefore, energy transmitters need to be scheduled with a guaranteed coverage area and duration to ensure efficient dedicated energy harvesting.

7.4.5 Energy Routing

The multiple path energy routing method collects distributed RF energy from numerous sources using RF energy routers. This allows energy to be sent to IoT devices via alternative path. In energy routing, relay nodes are placed near IoT devices as part of the multihop energy transmission concept. The proximity of the relay node to the IoT devices reduces path loss and improves the efficiency of RF to DC conversion.

7.5 Green Energy

Any form of energy produced from renewable energy sources is considered to be "green energy." Common sources of green energy include solar energy, wind power, geothermal energy, biomass, and hydroelectric power. Unlike fossil fuels such as coal or natural gas, which take millions of years to form, green energy sources are typically supplied by nature. Green sources, in contrast, avoid mining and drilling operations that harm ecosystems. Green energy is crucial for the environment as it provides environmentally preferable alternatives to the harmful effects of fossil fuels. It is derived from natural resources, is typically clean, renewable, and emits little to no greenhouse gases. It is also often readily available. This is beneficial for the environment as well as for the health of the individuals and animals who breathe the air. Green energy sources, are being produced locally and less susceptible to geopolitical crisis, price surges, or supply chain disruptions, result in more stable energy prices. Green energy is an affordable option to meet the global energy demands.

7.5 Green Energy

In the future, green energy has the potential to displace fossil fuels, requiring the implementation of various production techniques. Geothermal energy, for instance, is most effective in areas where accessing this resource is straightforward, while wind or solar energy may be more suitable in other regions. With green energy presenting a better alternative to many energy sources, it appears inevitable that it will become an integral part of the global future. These energy sources are easily replenished, beneficial for the environment, creating jobs, and expected to become economically viable as advancements continue. The fact is that fossil fuels must be phased out as they cannot sustainably meet our energy demands. By developing a range of green energy alternatives, we establish a fully sustainable future for energy supply without harming the environment in which we all reside. TWI has been involved in various green energy projects for decades, accumulating expertise in these fields and providing solutions to industrial members, including advancements in renewable energy [69].

7.6 Conceptual Framework of Proposed Energy-Based Quantum IoT Network Architecture for Sustainable Smart City

A new architecture is proposed for the conceptual framework of an energy-based Quantum IoT network architecture for a sustainable smart city, as depicted in Fig. 7.7. It consists of four layers, which have their own roles in the network as stated further.

7.6.1 Physical Layer

This layer will include all quantum hardware and optical fibers as well as all IoT devices in a network. This layer does not change from the current IoT network's current layer. The devices will send conventional bits to the gateway as usual, but the gateway will then transfer those bits to the quantum server, which will further turn them into qubits at the quantum network layer. Timing and synchronization will be handled by the physical layer hardware.

Sensors It is widely considered that sensors are essential to IoT technologies because of their ability to collect and transport data in real time [70–72]. As a vital component of the IoT, energy-related sensors are used for both cost- and energy-saving purposes. The use of sensors in the energy sector enhances the proportion of renewable energy sources, making the target of achieving optimal energy usage

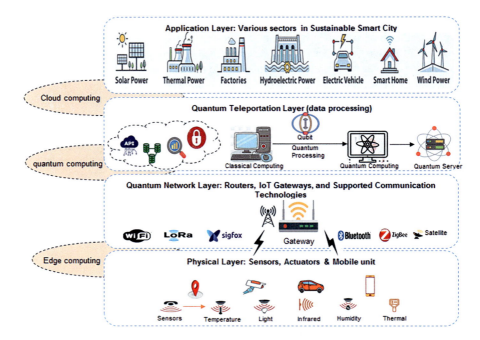

Fig. 7.7 Proposed energy-based quantum IoT architecture for sustainable smart city

more feasible. The most popular sensors are temperature sensors, humidity sensors, light sensors, passive infrared sensors, and proximity sensors. The energy generation and consumption in both phases are significantly influenced by temperature sensors. They are used to detect temperature differences in various environments, including the cooling and heating systems in energy production. Humidity sensors are used to measure the environment's moisture content and relative humidity. Humidity sensors are widely used in the energy sector, including in the production of wind energy, to measure and monitor moisture levels. Similar to this, light sensors are used in both residential and business appliances to determine how bright a light is [73]. To conserve energy, they turn on and off the lights automatically [74]. Passive infrared sensors, commonly referred to as movement sensors, are another variety of sensor. They are used, as their name suggests, to track the movement of things by detecting the infrared light radiation those objects emit when they are in a certain environment. These sensors are used in a various fields, including the energy sector, where they play a vital role in reducing building energy consumption by detecting and responding to movement [75].

Actuators Actuators differ from sensors include that it takes an electrical input and turn it into a specific kind of motion in order to carry out tasks in automation systems. Different types of actuators are used at different stages in the energy sector, for example, pneumatic actuators are frequently used as the final control element in power plant operations. Additionally, they are used to reduce energy waste when opening portals, tightening wind turbine brakes, and causing motions in solar tracking panels. For instance, the research in [76] shows how the IoT-based autonomous intelligent process benefits from remote sensors and actuators. The suggested solution lowers the energy consumption of devices operating in IoT systems.

7.6.2 Quantum Network Layer

This layer controls the transmission of both quantum and conventional data over the internet and local network. It includes a quantum device that handles all operations related to quantum data and performs conversions between conventional bits and quantum bits. The existing hardware acts as a gateway, similar to a router, facilitating the transfer of conventional bits between the physical layer to the quantum server in the quantum network layer.

Supported Communication Technologies An essential component of the IoT functional mechanism, the wireless communication framework enables end to end communication between IoT devices by connecting sensor devices to IoT gateways. The selection of a wireless standard for each application depends on specific requirements such as communication range, data transfer capacity, and power usage. The development of wireless frameworks encompasses a variety of wireless standards. For instance, many renewable energy sources, such as wind and solar power plants, are often situated in remote regions, posing challenges for reliable IoT communication. In such locations, it is crucial to choose a communication technology that can ensure a consistent link and efficiently deliver real-time data transfer for IoT frameworks. Examples of communication technologies include Wi-Fi, bluetooth, Zigbee, long range (LoRa), Sigfox, LTE-M, and bluetooth low energy (BLE). The application of short-range wireless communication technologies, such as Wi-Fi, narrowband IoT (NB-IoT), ZigBee, BLE, as well as LPWAN technologies like LoRa, Sigfox, and LTE-M running in the unlicensed band [77–81]. However, it is worth noting that Wi-Fi technologies are considered less efficient compared to other comparable technologies due to their higher energy consumption [71].

LPWAN technologies are expected to offer more energy-efficient solutions than Wi-Fi technologies in the near future [82]. In a similar manner, [83] demonstrated how emerging LPWAN technologies enable the establishment of reliable, simple, low power, and long-term effective energy management solutions. These technologies not only consume less energy but also have lower installation and operating expenses compared to alternative technologies [84, 85]. Another important technology is called Zigbee, which is utilized to establish secure network connectivity. Zigbee shares similarities with BLE in terms of easy installation, affordability, low data rate transmission, and reliable networks for low power devices [86, 87].

Zigbee is commonly used in IoT devices for lighting frameworks, smart grids, home automation frameworks, and industrial automation.

Long-term evolution for machine-type communications (LTE-M), Sigfox, narrowband IoT, and LoRa are other examples of long-range communication technologies. Among these, LoRa is a widely used communication technology for IoT. LoRa provides low power and cost-effective connectivity solutions for IoT applications, with a range that extend up to approximately 50 kilometers [88]. Several studies, such as those conducted by [89–91], explore the applications of these technologies in smart homes and industrial HVAC systems. Sigfox and LoRa share similarities in terms of their utility, energy consumption, installation costs, and coverage areas. However, Sigfox has significantly slower data transfer speeds compared to LoRa. Another cost-effective option with extended battery life and the ability to update the battery is NB-IoT. Additionally, satellite communication technology has emerged as a new communication technology that supports low data rate applications in machine-to-machine (M2M) communications [92]. Satellite technology is particularly suitable for providing IoT device backup in remote locations. Research has also presented the integration of satellite communication technology into the smart grid, solar systems, and wind turbine power generation [93, 94].

7.6.3 Quantum Teleportation Layer

The Quantum transportation layer consists solely of quantum repeaters, whose primary function is to preserve the integrity of qubits states and transmit them to the next hop. Quantum repeaters are employed to prevent the loss of entanglement in qubits during long distances travel.

7.6.4 Application Layer

The application layer serves as the connection point between the client and the information in the system. Qubits facilitate communication between the client and the next layer through command exchange. This layer provides applications with access to both classical and quantum capabilities of the network.

7.7 Enabling Technologies in Energy Sector

A large portion of the energy is wasted during energy consumption, and experts believe that efficient energy monitoring and smart energy solutions reduces energy demand. Smart energy systems not only reduce demand but also increase energy efficiency; thus smart solutions based on emerging technologies are imperative to

7.7 Enabling Technologies in Energy Sector

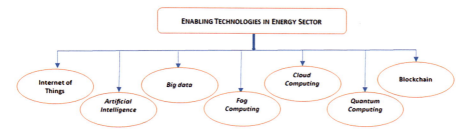

Fig. 7.8 Enabling technologies in energy sector

decrease energy demand and increase efficient energy management system. Emerging technologies such as the IoT, AI, big data, computing technologies, and blockchain are seen as major enablers in energy sector as shown in Fig. 7.8. Research studies show that disruptive technologies have a greater role in the development of efficient energy management system. Among disruptive technologies, IoT has revealed the potential to assist the development of efficient energy management system. In [95], IoT is the backbone of current and future energy management system, and at any level of the supply chain, it can monitor and increase the awareness about energy performance and real-time energy consumption. Based on the discussion, the current chapter presents key literature on IoT and related technologies, their applications in energy generation, intelligent transport, smart factories, smart buildings, smart grids, and major industrial application areas.

7.7.1 Internet of Things

IoT is referred to as an emerging innovation that utilizes the internet to establish connections among physical devices or "things" [96]. By utilizing sensors and correspondence systems effectively, these devices provide valuable information and services to people. For instance, implementing smart energy management systems in buildings enables efficient energy utilization, resulting in cost reduction [97]. IoT finds extensive applications in various industries such as manufacturing, services, and construction [98]. It is also widely used in environmental management, healthcare systems and services, energy management in buildings, and robot-based management [99–101]. Since IoT devices generate an enormous amount of data, efficient data storage systems are necessary. Different data storage mechanism are employed, such as a cloud servers or local storage within the IoT network. The data collected through IoT devices plays a crucial role in providing insightful information through data analytics. Real-time analysis often requires cloud servers for visualization or stream analytics. IoT is a paradigm in which items and components of a framework that are prepared with sensors, actuators, and processors can link with one another to offer significant types of service. Initially, IoT systems collect a massive amount of data through sensors and other technologies, which is then transformed into meaningful information using various analysis techniques. After analysis, the information is sent back to the actuator. This process involves the presence of multiple actuators, data, and computing devices.

7.7.2 Artificial Intelligence

AI is widely regarded as one of the most impactful disruptive technologies of the current era. AI transforms mechanical machines into intelligent devices by simulating human brain functions and utilizing cognitive patterns. Advancements in AI aim to imitate and develop human-like cognitive capabilities, ultimately enabling machines to function similarly to the human brain [102, 103]. The availability of a large amount of data presents significant opportunities for businesses to improve decision-making efficiency. However, handling such vast and complex data poses challenges for simple computational techniques to process and extract useful information for decision-makers. In such a situation, AI employs two main computational mechanisms to process, classify, and generate key information from big data. The AI paradigm is divided into two fundamental categories: Artificial Neural Networks (ANNs) and Conventional Artificial Intelligence (CAI). CAI, compared to ANNs, operates and responds in more general computing style by following predetermined rules and knowledge provided by human experts.

In AI, machine learning is considered one of the main categories and enhances computational powers of a system through learning algorithms. In [104] explains that there are three main types of learning algorithms: supervised, unsupervised, and hybrid learning algorithms. Each algorithm has specific conditions and applications based on the requirements and purpose of using AI. Despite the availability of multiple models and algorithms related to machine learning, studies and implementation of machine learning are in their initial stages.

7.7.3 Cloud/Fog/Edge Computing

In the energy sector, this huge amount of data is utilized to enhance energy efficiency, reduce consumption, and drive development [105]. However, as data is collected from various sources, it includes a large volume of raw data, necessitating the use of sophisticated computing techniques to classify highly relevant and irrelevant data sets within big data [106]. Cloud and fog computing are two major computing mechanisms known for handling and processing big data.

Cloud computing is TCP/IP-based technology that has seen significant development and integration of computer technologies, including fast microprocessor, large memory, high speed networks, and reliable system architecture [107]. The

7.7 Enabling Technologies in Energy Sector

infrastructure of cloud computing is composed of five layers, i.e., clients, applications, platform, infrastructure, and servers [107]. These layers encompass hardware, software, and services. Internet and communication protocols also play a crucial role in the functioning of cloud computing. It is a highly sophisticated technique with the potential to process and analyze IoT-based big data. In cloud computing, users interact with the service layer through the internet, and secure access is provided to users to access the services. However, the hardware of cloud servers is located in large data centers at different locations than the users [108]. Many organizations are adopting cloud services due to its advantages, such as reducing the hardware expenses, providing extensive data storage, offering secure multi-layered architecture, and facilitating accessibility from multiple geographical locations [109]. Despite the popularity and the opportunities it provides for data computation and analytics, cloud computing also has certain limitations, such as delays in accessing the server and bandwidth issues.

In context to smart cities, the need for a large-scale data center to accommodate and process energy-related big data arises. However, the increasing number of physical machines in data centers leads to disadvantages in terms of initial installation costs, operating costs, cooling costs, and more. To address these challenges, cloud computing has been proposed as a solution. Cloud computing enables the on-demand availability of virtual machines, meaning that data processing is performed not on customers' physical computers but on virtual machines connected to a large-scale cloud infrastructure. This introduction of cloud computing helps reduce costs associated with individual purchases, installation or updates, as well as time and labor costs. The existing data storage system often have limitations in terms of slow data processing capacity and finite data storage capabilities. In contrast, cloud computing offers fast data processing capacity and handles big data more efficiently.

Fog computing is a decentralized and extended form of cloud computing that serves as an intermediary between a cloud server and client hardware. In the context of IoT, collected data is processed locally instead of being sent to the server, resulting in more reliable and faster data responses while reducing network traffic. These issues, such as delays and bandwidth limitations, often hinder the efficiency of the system. Therefore, a decentralized computing approach is necessary to overcame these challenges. In this context, fog computing can be utilized as a complementary method to cloud computing.

Edge computing technologies are relatively new and challenging to find literature specifically focusing on these techniques in smart home technologies. The primary objective of these techniques is to minimize the response time for requests directed towards a smart grid. In this context, researchers have explored the utilization of fog computing layers to efficiently manage the requests from smart home consumers to the smart grids. It is a growing field in smart home technology, which aims to shift the computing power towards the fog layers in order to mitigate delays and improve response time.

7.7.4 Quantum Computing

The world's most powerful supercomputers and cloud data centers consume significant amounts of energy to solve various problems. In comparison, quantum computers are expected to be more energy efficient when performing specific tasks. Quantum computers can reliably perform extensive calculations using less energy, resulting in reduced costs and carbon emissions. Unlike classical computers that use binary bits (0 or 1) to encode information, quantum computing utilizes qubits, which represent both 0s and 1s simultaneously. This property of quantum computing allows for the identification of optimal solutions while consuming less energy. The reduced energy consumption is attributed to quantum processors operating at shallow temperature, and being superconducting, thereby generating no resistance or heat. Hybrid applications typically consist of two components: high energy and low energy.

Quantum computing executes the high energy portion, while classical computing handles the low energy part through the cloud. To address these types of problems, the adoption of hybrid computing, which combines quantum and classical computing, is crucial in significantly reducing energy usage and costs. However, more research and development work is needed before hybrid computing can effectively solve today's most challenging business problems. Additionally, machine learning techniques play a crucial role in predicting the energy demand of quantum computers, considering both renewable and non-renewable energy sources. Furthermore, the utilization of effective data analytics techniques enables accurate predictive analytics for energy consumption. Therefore, there is a need to develop the energy-efficient quantum data centers to optimize energy utilization.

7.7.5 Big Data

IoT technologies play a significant role in collecting a vast amount of data often referred to as big data; which is then utilized for making key decisions in business operations and processes. With the increasing importance of ICT in smart cities and the expanding scope of smart metering, there is a growing focus on the processing and analysis methods for big data. Consequently, numerous studies are now exploring methodologies for data analysis, such as AI, machine learning, deep learning, and Q-learning. In recent years, big data analysis has emerged as a vital tool not only for acquiring information on historic trends but also in formulating a future energy plan for smart cities through forecasting. For example, district heating systems exhibit different energy response patterns based on the characteristics of individual building. Therefore, it is necessary to incorporate these differences into thermal load forecasting to ensure accurate and effective energy planning.

7.7.6 Blockchain

Blockchain technology is a decentralized and transparent system for storing and transmitting operational information securely, without the need for a central control authority. It utilizes a chain of blocks to create a database that maintains a complete history of all transactions conducted by its users since its inception. With its numerous advantages, blockchain technology begun to expand into various industries, including energy, automotive, healthcare, and others. The key advantages of blockchain technology include transparency, decentralization, immutability, and security. In the energy sector, blockchain is considered a promising tool to record and facilitate transactions between generators and consumers of energy. It enables several use cases, such as peer-to-peer electricity trading, management of renewable energy certificates, grid management, and electric vehicle charging. It has the potential to revolutionize the energy industry by providing a more efficient and secure way to manage energy transactions and promote the adoption of renewable energy sources.

Blockchain technology is highly promising in the power sector, as evidenced by its relevant use cases. As a matter of fact, the combination of new initiatives using various initiatives leveraging blockchain technology have emerged to enhance the efficiency and transparency of the sector, facilitate the integration of distributed energy sources, and reduce costs. These initiatives are opening up avenues for a promising future for blockchain in the power sector.

7.8 Open Research Challenges and Research Directions

With the increase in consumer demands, energy management has become an essential necessity worldwide. The global warming issue and climate change pose a major threat to the future generations. As a consequence of increasing energy consumption, greenhouse gases have been released. However, Cisco estimates that there will be over fifty billion connected smart devices in the near future. The increasing the number of IoT devices will result in significant energy consumption. Therefore, effective energy management for IoT devices is critical for the long-term realization of smart cities. The following are a few instances of how excellent management reduce energy use, as depicted in Fig. 7.9.

- *Home Appliance/Domestic Services/Rural and Sub-urban Areas*

Generally, household appliances and domestic services in rural/sub-urban areas consume the most energy. Light sources, temperature control, and thermal management are part of demand management, which allows residential units to be customized in terms of energy consumption. Alternatively, intelligent activity management ensures efficient and effective energy management.

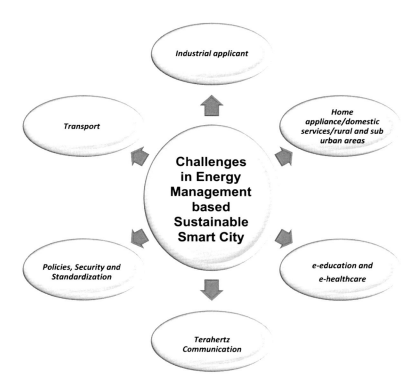

Fig. 7.9 Challenges in energy management-based sustainable smart city

- *e-Education and e-Healthcare*

  There are challenges in dematerializing education and health services due to their important role in smart cities. However, it is possible to reduce energy usage by demobilizing services. For example, remote healthcare using smart phones and sensor technology saves energy, as does distance education.

- *Industrial Applicant*

  A system that uses the IoT are utilized to make better decisions about food availability. Similarly, food transportation is improved with intelligent means of transportation. IoT devices usually run on batteries and have limited storage capacity. As a result of these sensor limitations, implementation of IoT systems with prolonged network life is challenging. An optimized energy-efficient architecture is essential for making the most of the limited sensor resources. It will reduce energy consumption and also ensure a minimal level of quality of service.

- *Transport*

  As part of transportation, energy is primarily used for public transit service, commuting in private vehicles to and from work, and other activities, and more. The high consumption of energy by public transportation and personal vehicles constitutes a significant source of pollution in cities. IoT-enabled technologies are utilized

7.8 Open Research Challenges and Research Directions

for energy management, including transportation planning, traffic engineering, and smart parking. They help reduce both energy consumption and greenhouse gas emissions significantly.

- *Terahertz Communication*

One of the primary challenges that IoE faces in maintaining device connections in smart cities is the limited spectrum availability compared to the increasing demand for higher data rates. Traditional methods for increasing data speed, such as increasing bandwidth or spectral efficiency, have been widely used. However, relying solely on spectral efficiency will not be sufficient to meet the future data traffic demands. New enabling technologies like terahertz (THz) transmission will be necessary to address the massive bandwidth requirements of IoE for smart cities. By utilizing the THz spectrum, which encompasses the frequency range of 0.3–10 THz, it becomes possible to achieve higher peak data rates and accommodate the growing needs of IoE applications.

- *Policies, Security, and Standardization*

The IoE offers numerous benefits, including adaptable technological platforms. However, it is crucial to emphasize the importance of coordination and cooperation among technologies, networks, and entities. Both producers and consumers face the challenges related to interconnections, policies, security, and standardization. Innovative theoretical models and protocol standards are required for effective energy management systems. To overcome the challenges associated with the IoE, it is necessary to enhance energy production efficiency and ensure are liable supply of energy from all sources. This, in turn, leads to lower energy development and production costs across the industry, making energy more accessible to consumers.

For the implementation of sustainable smart cities and the advancement of energy conservation system in the future, it is imperative to develop new energy infrastructure. Future research challenges in this domain include the development of real-time energy monitoring systems, which can be effectively addressed through the utilization of AI technology. In addition, the application of intelligent energy management technologies plays a crucial role in effectively managing energy performance. The development of big-data analytics technologies are developed for the intelligent energy management. Moreover, the implementation of integrated energy network technologies is vital for ensuring sustainability in smart cities. With the advancement of new technologies, the concept of the "IoE" is gaining traction. This concept involves upgrading and automating energy infrastructure through the combined use of energy conservation systems and IoT technologies. From a city perspective, the integration of IoT technology and energy infrastructure is crucial. The energy network plays a vital role in enabling flexible energy trading between distributed energy generation systems and ensuring efficient management of energy peak loads through demand response mechanisms. By utilizing an IoT-based integrated energy network system, various emerging technologies for the effective operation of the advanced energy conservation system can be applied to sustainable smart cities. Moreover, if the usability and feasibility of the integrated energy network system are verified with different infrastructures, it can be implemented worldwide to create sustainable smart cities in the future.

7.9 Summary

In smart cities, energy management is an unavoidable challenge due to increasing urbanization. This chapter discusses energy management in smart cities and its classification in IoE-enabled cities. We explore energy management in relation to energy-efficient technologies and operations that enable energy harvesting. Smart cities encounter various energy-efficient solutions and face challenges in energy harvesting. IoE solutions with software-defined features and energy-efficient techniques have the potential to provide scalable services. Ensuring the energy efficiency and complexity of security protocols is crucial for their practical application in the IoE, especially for energy-constrained smart devices. Researching power-efficient security protocols for such devices is therefore essential. Additionally, fog computing holds promise for energy savings in IoE applications.

References

1. P. Mishra, P. Thakur, G. Singh, Enabling technologies for IoT based smart city, in *Proceedings of 2021 Sixth International Conference on Image Information Processing (ICIIP)*, vol. 6, (IEEE, 2021), pp. 99–104
2. United Nations, *World Population Prospects 2019: Highlights* (United Nations Publication, 2019) Accessed on 8 Mar 2023
3. International Energy Agency. World energy outlook 2013. Accessed on 9 Mar 2023
4. Available online https://sdgs.un.org/goals/goal7. Accessed on 8 Mar 2023
5. P. Mishra, P. Thakur, G. Singh, Sustainable smart city to society 5.0: State-of-the-art and research challenges. SAIEE Afr. Res. J. **113**(4), 152–164 (2022)
6. Available online https://unstats.un.org/sdgs/. Accessed on 8 Mar 2023
7. D. Connolly et al., Heat roadmap Europe: Combining district heating with heat savings to decarbonise the EU energy system. Energy Policy **65**, 475–489 (2014)
8. A. Razmjoo, R. Shirmohamadi, A. Davarpanah, F. Pourfayaz, A. Aslani, Stand-alone hybrid energy systems for remote area power generation. Energy Rep. **5**, 231–241 (2019)
9. C. Lim, K.J. Kim, P.P. Maglio, Smart cities with big data: Reference models, challenges, and considerations. Cities **82**, 86–99 (2018)
10. N. Hossein Motlagh, M. Mohammadrezaei, Internet of things (IoT) and the energy sector. Energies **13**(2), 494/1–494/27 (2020)
11. S. Optimal, M. Narodoslawsky, Renewable energy systems for smart cities. Comp. Aided Chem. Eng. **33**, 1849–1854 (2014)
12. W. Ejaz, A. Anpalagan, Internet of things enabled electric vehicles in smart cities, in *Internet of Things for Smart Cities: Technologies, Big Data and Security*, (Springer, Cham, 2019), pp. 39–46
13. B. Hu, Y. Feng, J. Sun, Y. Gao, J. Tan, Driving preference analysis and electricity pricing strategy comparison for electric vehicles in smart city. Inf. Sci. **504**, 202–220 (2019)
14. X. Xu, D. Niu, Y. Li, L. Sun, Optimal pricing strategy of electric vehicle charging station for promoting green behavior based on time and space dimensions. J. Adv. Transp. **2020**, 1–16 (2020)
15. F. Pinna, F. Masala, C. Garau, Urban policies and mobility trends in Italian smart cities. Sustainability **9**(4), 494 (2017)

16. B. Jan, H. Farman, M. Khan, S.H. Ahmad, An adaptive energy efficient scheme for energy constrained wireless sensor networks, in *34th ACM/SIGAPP Symposium on Applied Computing Limassol, Cyprus*, (8 April 2019), pp. 2391–2398
17. T. Ozawa et al., Smart cities and energy management. Fujitsu Sci. Tech. J. **50**(2), 49–57 (2015)
18. D. Bonino et al., ALMANAC: Internet of things for smart cities, in *FiCloud 2015, 3rd IEEE International Conference on Future Internet of Things and Cloud*, (August 2015), pp. 309–316
19. H.C. Pohls et al., RERUM: Building a reliable iot upon privacy-and security-enabled smart objects, in *IEEE Wireless Communications and Networking Conference Workshop*, (Istanbul, Turkey, April 2014), pp. 122–127
20. C. Subhes Bhattacharyya, G.R. Timilsina, A review of energy system models. Int. J. Energy Sect. Manag. **4**(4), 494–518 (2010)
21. A. Soroudi, E. Mehdi, A distribution network expansion planning model considering distributed generation options and techo-economical issues. Energy **35**(8), 3364–3374 (2010)
22. S. Karnouskos, T.N. de Holanda, Simulation of a smart grid city with software agents, in *Third UK Sim European Symposium on Computer Modeling and Simulation*, (IEEE, 2009), pp. 424–429
23. A.M. Foley, B.P.O. Gallachoir, J. Hur, R. Baldick, E.J. McKeogh, A strategic review of electricity systems models. Energy **35**(12), 4522–4530 (2010)
24. S. Malekpour, R.R. Brown, F.J. de Haan, Strategic planning of urban infrastructure for environmental sustainability: Understanding the past to intervene for the future. Cities **46**, 67–75 (2015)
25. I.-A. Yeo, J.-J. Yee, A proposal for a site location planning model of environmentally friendly urban energy supply plants using an environment and energy geographical information system (E-GIS) database (DB) and an artificial neural network (ANN). Appl. Energy **119**, 99–117 (2014)
26. J. Lenhart, B. van Vliet, A.P.J. Mol, New roles for local authorities in a time of climate change: The Rotterdam Energy Approach and Planning as a case of urban symbiosis. J. Clean. Prod. **107**, 593–601 (2015)
27. I. Maric, M. Pucar, B. Kovacevic, Reducing the impact of climate change by applying information technologies and measures for improving energy efficiency in urban planning. Energ. Buildings **115**, 102–111 (2016)
28. K. Meng, Z.Y. Dong, Z. Xu, Coordinated dispatch of virtual energy storage systems in smart distribution networks for loading management. IEEE Trans. Syst. Man Cybern. Syst. **49**(4), 776–786 (2017)
29. M. Collotta, G. Pau, An innovative approach for forecasting of energy requirements to improve a smart home management system based on BLE. IEEE Trans. Green Commun. Netw. **1**(1), 112–120 (2017)
30. S. Chen, T. Liu, F. Gao, J. Ji, X. Zhanbo, B. Qian, W. Hongyu, X. Guan, Butler, not servant: A human-centric smart home energy management system. IEEE Commun. Mag. **55**(2), 27–33 (2017)
31. S.L. Arun, M.P. Selvan, Intelligent residential energy management system for dynamic demand response in smart buildings. IEEE Syst. J. **12**(2), 1329–1340 (2017)
32. B.V. Solanki, A. Raghurajan, K. Bhattacharya, Including smart loads for optimal demand response in integrated energy management systems for isolated microgrids. IEEE Trans. Smart Grid **8**(4), 1739–1748 (2015)
33. H.S.V.S. Kumar Nunna, S. Battula, S. Doolla, Energy management in smart distribution systems with vehicle-to-grid integrated microgrids. IEEE Trans. Smart Grid **9**(5), 4004–4016 (2016)
34. N. Kumar, S. Zeadally, S.C. Misra, Mobile cloud networking for efficient energy management in smart grid cyber-physical systems. IEEE Wirel. Commun. **23**(5), 100–108 (2016)

35. Y. Xin Wang, T.C. Zhang, G.B. Giannakis, Dynamic energy management for smart-grid-powered coordinated multipoint systems. IEEE J. Sel. Areas Commun. **34**(5), 1348–1359 (2016)
36. G.K. Venayagamoorthy, R.K. Sharma, P.K. Gautam, A. Ahmadi, Dynamic energy management system for a smart microgrid. IEEE Trans. Neural Netw. Learn. Syst. **27**(8), 1643–1656 (2016)
37. A. Mondal, S. Misra, M.S. Obaidat, Distributed home energy management system with storage in smart grid using game theory. IEEE Syst. J. **11**(3), 1857–1866 (2015)
38. K. Wang, Z. Ouyang, R. Krishnan, L. Shu, L. He, A game theory-based energy management system using price elasticity for smart grids. IEEE Trans. Industr. Inform. **11**(6), 1607–1616 (2015)
39. Available online https://www.cisco.com/c/dam/en_us/about/business-insights/docs/ioe-value-at-stake-public-sector-analysis-faq.pdf. Accessed on 18 Mar 2023
40. Y. Shahzad, H. Javed, H. Farman, J. Ahmad, B. Jan, M. Zubair, Internet of energy: Opportunities, applications, architectures and challenges in smart industries. Comput. Electr. Eng. **86**, 106739/1–106739/14 (2020)
41. M. Curiale, From smart grids to smart city, in *Proceedings of the 2014 Saudi Arabia Smart Grid Conference*, (SASG, Jeddah, Saudi Arabia, December 2014), pp. 14–17
42. K. Parham, M. Farmanbar, C. Rong, O. Arild, Assessing the importance of energy management in smart homes, in *Proceedings of the 4th International Conference on Viable Energy Trends (InVEnT-2019)*, (Istanbul, Turkey, April 2019), pp. 26–28
43. V.C. Gungor, D. Sahin, T. Kocak, S. Ergut, C. Buccella, C. Cecati, G.P. Hancke, Smart grid and smart homes: Key players and pilot projects. IEEE Ind. Electron. Mag. **6**, 18–34 (2012)
44. T. Chen, Smart grids, smart cities need better networks. IEEE Netw. **24**, 2–3 (2010)
45. M. Yun, B. Yuxin, Research on the architecture and key technology of internet of things (IoT) applied on smart grid, in *2010 International Conference on Advances in Energy Engineering, ICAEE*, (2010), pp. 69–72
46. Z. Xu, X. Li, The construction of interconnected communication system among smart grid and a variety of networks, in *Proceedings of Asia-Pacific Power and Energy Engineering Conference, APPEEC*, (2010), pp. 1–5
47. R.A. Leon, V. Vittal, G. Manimaran, Application of sensor network for secure electric energy infrastructure. IEEE Trans. Power Deliv. **22**(2), 1021–1028 (2007)
48. A. Meloni, P.A. Pegoraro, L. Atzori, A. Benigni, S. Sulis, Cloud-based IoT solution for state estimation in smart grids: Exploiting virtualization and edge-intelligence technologies. Comput. Netw. **130**, 156–165 (2018)
49. I. Farris, R. Girau, L. Militano, M. Nitti, L. Atzori, A. Iera, et al., Social virtual objects in the edge cloud. IEEE Cloud Comput. **2**(6), 20–28 (2015)
50. L. Atzori, A. Iera, G. Morabito, The internet of things: A survey. Comput. Netw. **54**(15), 2787–2805 (2010)
51. I. Cvetkovic, *Modeling, Analysis and Design of Renewable Energy Nanogrid* (Dissertation of Master of Sciences, Virginia Tech, EUA, 2010)
52. R.H. Lasseter, P. Paigi, Microgrid: A conceptual solution, in *Proceedings of Power Electronics Specialists Conference*, vol. 6, (2004), pp. 4285–4290
53. R. Lasseter, Microgrid and distributed generation. J. Energy Eng. (American Society of Civil Engineers) **133**(3), 144–149 (2007)
54. D. Dietrich, D. Bruckner, G. Zucker, P. Palensky, Communication and computation in buildings: A short introduction and overview. IEEE Trans. Ind. Electron. **57**, 3577–3584 (2010)
55. X. Yu, C. Cecati, T. Dillon, M.G. Simoes, The new frontier of smart grids. IEEE Ind. Electron. Mag. **5**, 49–63 (2011)
56. M. Liserre, T. Sauter, J.Y. Hung, Future energy systems: Integrating renewable energy sources into the smart power grid through industrial electronics. IEEE Ind. Electron. Mag. **4**, 18–37 (2010)

References

57. P. Palensky, D. Dietrich, Demand side management: Demand response, intelligent energy systems, and smart loads. IEEE Trans. Ind. Inform. **7**, 381–388 (2011)
58. M. Li, H. Xiao, W. Gao L. Li, Smart grid supports the future intelligent city development, in *Proceedings of the 28th Chinese Control and Decision Conference, CCDC*, (Yinchuan, China, 28–30 May 2016), pp. 6128–6131
59. Smart Systems Thinking for Comprehensive City Efficient Energy Planning. Available online http://www.smartsteep.eu. Accessed on 1 Jan 2023
60. Stepup, Energy Planning for Cities. Available online http://www.stepupsmartcities.eu/. Accessed on 24 Mar 2023
61. Planning for Energy Efficient Cities (PLEEC). Available online http://www.pleecproject.eu. Accessed on 13 Feb 2023
62. ZenN-ZenN, Nearly Zero Energy Neighbourhoods. Available online http://zenn-fp7.eu. Accessed on 7 Dec 2022
63. R2Cities: Residential Renovation Towards Nearly Zero Energy Cities. Available online http://r2cities.eu. Accessed on 7 Feb 2023
64. Ready, Resource Efficient Cities Implementing Advanced Smart City Solutions. Available online http://www.smartcity-ready.eu. Accessed on 16 Jan 2023
65. W. Ejaz et al., Energy and throughput efficient cooperative spectrum sensing in cognitive radio sensor networks. Trans. Emerg. Telecommun. Technol. **26**(7), 1019–1030 (2015)
66. P. Vlacheas et al., Enabling smart cities through a cognitive framework for the internet of things. IEEE Commun. Mag. **51**(6), 102–111 (2013)
67. F. Qayyum et al., Appliance scheduling optimization in smart home networks. IEEE Access **3**, 2176–2190 (2015)
68. V. Grigonis, M. Burinskiene, Information technologies in energy planning of cities and towns. J. Civ. Eng. Manag. **8**(3), 197–205 (2002)
69. https://www.twi-global.com/technical-knowledge/faqs/what-is-green-energy#TypesofGreenEnergy. Accessed on 24 Dec 2022
70. S.D.T. Kelly, N.K. Suryadevara, S.C. Mukhopadhyay, Towards the implementation of IoT for environmental condition monitoring in homes. IEEE Sensors J. **13**(10), 3846–3853 (2013)
71. S.A.A. Abir, A. Anwar, J. Choi, A.S.M. Kayes, Iot-enabled smart energy grid: Applications and challenges. IEEE Access **9**, 50961–50981 (2021)
72. T. Rault, A. Bouabdallah, Y. Challal, Energy efficiency in wireless sensor networks: A top-down survey. Comput. Netw. **67**, 104–122 (2014)
73. L. Perez-Lombard, J. Ortiz, C. Pout, A review on buildings energy consumption information. Energ. Buildings **40**(3), 394–398 (2008)
74. N.H. Motlagh, S.H. Khajavi, A. Jaribion, J. Holmstrom, An IoT-based automation system for older homes: A use case for lighting system, in *Proceedings of 2018 IEEE 11th Conference on Service-Oriented Computing and Applications (SOCA)*, (IEEE, 2018), pp. 1–6
75. W. Kim, K. Mechitov, J.Y. Choi, S. Ham, On target tracking with binary proximity sensors, in *Proceedings of IPSN 2005, Fourth International Symposium on Information Processing in Sensor Networks*, (IEEE, 2005), pp. 301–308
76. J. Blanco, A. Garcia, J.D.L. Morenas, Design and implementation of a wireless sensor and actuator network to support the intelligent control of efficient energy usage. Sensors **18**(6), 1892 (2018)
77. C. Eugenio, Manufacturing low-cost wifi-based electric energy meter, in *Proceedings of 2014 IEEE Central America and Panama Convention (CONCAPAN XXXIV)*, (IEEE, 2014), pp. 1–6
78. E. Rodriguez-Diaz, J.C. Vasquez, J.M. Guerrero, Intelligent DC homes in future sustainable energy systems: When efficiency and intelligence work together. IEEE Consum. Electron. Mag. **5**(1), 74–80 (2015)
79. A. Karthika, K.R. Valli, R. Srinidhi, K. Vasanth, Automation of energy meter and building a network using IoT, in *Proceedings of 2019 5th International Conference on Advanced Computing & Communication Systems (ICACCS)*, (IEEE, 2019), pp. 339–341

80. T. Lee, S. Jeon, D. Kang, L.W. Park, S. Park, Design and implementation of intelligent HVAC system based on IoT and Bigdata platform, in *Proceedings of 2017 IEEE International Conference on Consumer Electronics (ICCE)*, (IEEE, 2017), pp. 398–399
81. Y.T. Lee, W.H. Hsiao, C.M. Huang, T.C. Seng-cho, An integrated cloud-based smart home management system with community hierarchy. IEEE Trans. Consum. Electron. **62**(1), 1–9 (2016)
82. Y. Kabalci, E. Kabalci, S. Padmanaban, J.B. Holm-Nielsen, F. Blaabjerg, Internet of things applications as energy internet in smart grids and smart environments. Electronics **8**(9), 972/1–972/16 (2019)
83. S. Jain, M. Pradish, A. Paventhan, M. Saravanan, A. Das, Smart energy metering using LPWAN IoT technology, in *Proceedings of ISGW 2017: Compendium of Technical Papers*, (Springer, Singapore, 2018), pp. 19–28
84. M. Choi, W.K. Park, I. Lee, Smart office energy management system using bluetooth low energy-based beacons and a mobile app, in *Proceedings of 2015 IEEE International Conference on Consumer Electronics (ICCE)*, (IEEE, 2015), pp. 501–502
85. M. Collotta, G. Pau, A novel energy management approach for smart homes using bluetooth low energy. IEEE J. Sel. Areas Commun. **33**(12), 2988–2996 (2015)
86. W.C. Craig, *Zigbee: Wireless Control That Simply Works* (Zigbee Alliance ZigBee Alliance, 2004)
87. I. Froiz-Miguez, T.M. Fernandez-Carames, P. Fraga-Lamas, L. Castedo, Design, implementation and practical evaluation of an IoT home automation system for fog computing applications based on MQTT and ZigBee-WiFi sensor nodes. Sensors **18**(8), 2660/1–2660/42 (2018)
88. A. Augustin, J. Yi, T. Clausen, W.M. Townsley, A study of LoRa: Long range & low power networks for the internet of things. Sensors **16**(9), 1466/1–1466/18 (2016)
89. B. Mataloto, J.C. Ferreira, N. Cruz, LoBEMS—IoT for building and energy management systems. Electronics **8**(7), 763/1–763/27 (2019)
90. A. Javed, H. Larijani, A. Wixted, Improving energy consumption of a commercial building with IoT and machine learning. IT Prof. **20**(5), 30–38 (2018)
91. J.C. Ferreira, J.A. Afonso, V. Monteiro, J.L. Afonso, An energy management platform for public buildings. Electronics **7**(11), 294/1–294/13 (2018)
92. J. Wei, J. Han, S. Cao, Satellite IoT edge intelligent computing: A research on architecture. Electronics **8**(11), 1247/1–1247/16 (2019)
93. K. Sohraby, D. Minoli, B. Occhiogrosso, W. Wang, A review of wireless and satellite-based m2m/iot services in support of smart grids. Mob. Netw. Appl. **23**(4), 881–895 (2018)
94. M. De Sanctis, E. Cianca, G. Araniti, I. Bisio, R. Prasad, Satellite communications supporting internet of remote things. IEEE Internet Things J. **3**(1), 113–123 (2015)
95. Y.S. Tan, Y.T. Ng, J.S.C. Low, Internet-of-things enabled real-time monitoring of energy efficiency on manufacturing shop floors. Procedia CIRP **61**, 376–381 (2017)
96. K. Haseeb, A. Almogren, N. Islam, I. Ud Din, Z. Jan, An energy-efficient and secure routing protocol for intrusion avoidance in IoT-based WSN. Energies **12**(21), 4174/1–4174/18 (2019)
97. A. Zouinkhi, H. Ayadi, T. Val, B. Boussaid, M.N. Abdelkrim, Auto-management of energy in IoT networks. Int. J. Commun. Syst. **33**(1), e4168/1–e4168/16 (2020)
98. J. Holler, V. Tsiatsis, C. Mulligan, S. Avesand, S. Karnouskos, D. Boyle, *From Machine-to-Machine to the Internet of Things: Introduction to a New Age of Intelligence* (Elsevier, Waltham, 2015)
99. M.U. Farooq, M. Waseem, S. Mazhar, A. Khairi, T. Kamal, A review on internet of things (IoT). Int. J. Comput. Appl. **113**(1), 1–7 (2015)
100. T.K. Hui, R.S. Sherratt, D.D. Sanchez, Major requirements for building smart homes in smart cities based on internet of things technologies. Futur. Gener. Comput. Syst. **76**, 358–369 (2017)
101. N.H. Motlagh, M. Bagaa, T. Taleb, Energy and delay aware task assignment mechanism for UAV-based IoT platform. IEEE Internet Things J. **6**(4), 6523–6536 (2019)
102. Z. Shi, *Advanced Artificial Intelligence*, vol 4 (World Scientific, Singapore, 2019)

References

103. R. Abduljabbar, H. Dia, S. Liyanage, S.A. Bagloee, Applications of artificial intelligence in transport: An overview. Sustainability **11**(1), 189/1–189/24 (2019)
104. A.K. Jain, M. Jianchang, K.M. Mohiuddin, Artificial neural networks: A tutorial. Computer **29**(3), 31–44 (1996)
105. A. Jaribion, S.H. Khajavi, N.H. Motlagh, J. Holmstrom, [WiP] a novel method for big data analytics and summarization based on fuzzy similarity measure, in *Proceedings of 2018 IEEE 11th Conference on Service-Oriented Computing and Applications (SOCA)*, (IEEE, 2018), pp. 221–226
106. I. Stojmenovic, Machine-to-machine communications with in-network data aggregation, processing, and actuation for large-scale cyber-physical systems. IEEE Internet Things J. **1**(2), 122–128 (2014)
107. C. Gong, J. Liu, Q. Zhang, H. Chen, Z. Gong, The characteristics of cloud computing, in *Proceedings of 39th International Conference on Parallel Processing Workshops*, (2010), pp. 275–279
108. M. Armbrust, A. Fox, R. Griffith, A.D. Joseph, R. Katz, A. Konwinski, G. Lee, D. Patterson, A. Rabkin, I. Stoica, M. Zaharia, A view of cloud computing. Commun. ACM **53**(4), 50–58 (2010)
109. I. Foster, Y. Zhao, I. Raicu, S. Lu, Cloud computing and grid computing 360-degree compared, in *2008 Grid Computing Environments Workshop*, (2008), pp. 1–10

Chapter 8
Internet of Vehicles for Sustainable Smart Cities

8.1 Introduction

The concept of smart cities refers to an area of the city where innovation is fostered through the use of digital networks and applications [1]. This concept emphasizes utilizing technology to increase efficiency and create a more sustainable environment. It also aims to improve the quality of life of citizens and ensure the fulfillment of their basic needs. The term smart city is often used to describe cities that are sustainable, digitally connected [2], and driven by cutting-edge technology and data-driven solutions. These cities address challenges such as traffic congestion, pollution, energy efficiency, public safety, and economic growth. They strive to create a more liveable, vibrant, and interconnected environment for their citizens. Urbanization and a growing urban population present challenges that are addressed by transforming a city into a smart environment. A smart city is an urban area that provides sustainable economic growth and a high quality of life. The utilization of intelligent solutions such as traffic congestion avoidance [3–5] and industrial control systems is one way to make urbanization sustainable in today's world. By implementing these solutions, cities are able to use resources more efficiently, reduce pollution, increase safety, and provide citizens with better services and information. In addition to attracting more businesses and investment, smart cities offer citizens greater job opportunities and a better quality of life. In a smart city, technology is intelligently utilized to enhance living, working, commuting, and sharing information [6]. As part of the Internet of Things (IoT) concept, next-generation vehicles incorporate advanced sensing, communication, and social capabilities. Transforming smart cities into a reality necessitates cars providing wireless sensing and communication capabilities facilitate data access.

In the modern society and the automotive industry, communication and information technology are driving forces. Since the advent of mobile communications two decades ago, citizens lives have been transformed by the ability to exchange

© The Author(s), under exclusive license to Springer Nature Switzerland AG 2023
P. Mishra, G. Singh, *Sustainable Smart Cities*,
https://doi.org/10.1007/978-3-031-33354-5_8

information anywhere and at any time. It is expected that mobile communications systems will soon become a reality in vehicles. Around the world, significant resources and efforts from industries, universities, and governments are dedicated to developing road transport infrastructure and vehicles. It is increasingly common for vehicles to be equipped with embedded devices and features, including sensors, cameras, computers, and communications tools. These features impact existing systems by enabling real-time data transmissions, interpretations, and decision-making to assist drivers and devices. Vehicles have become relevant tools for smart cities as they have the ability to sense objects and events in the environment and respond appropriately. They not only assist in vehicle traffic management but also serve as a means of capturing real-time, relevant information for resource management. The development of an intelligent city involves careful consideration of mobility and management, such as traffic, surveillance, natural disasters, and sustainability.

These types of solutions require the collection and dissemination of numerous urban data streams, which, in turn, necessitate integrated, heterogeneous, and intelligent wireless communication infrastructures. In smart city scenario, various physical objects, or "smart" objects, interact with each other. The IoT refers to these interconnected and interoperable smart objects, which work together to create a safe and smart environment. Within the realm of IoT, many of these objects will be connected cars or vehicles that communicate wirelessly with a variety of internet-connected devices, both within and outside the vehicle. The Internet of Vehicles (IoV) is customized type of IoT that enables unified transportation management and other smart city applications. Essentially, IoV is a convergence of IoT and mobile internet. IoV technology involves dynamic mobile communication systems or models that facilitate interactions between vehicles and other objects, including vehicle-to-vehicle (V2V), vehicle-to-road (V2R), vehicle-to-infrastructure (V2I), vehicle-to-building (V2B), vehicle-to-home (V2H), vehicle-to-everything (V2X), vehicle-to-grid (V2G), and vehicle-to-home (V2H) interactions as shown in Fig. 8.1. Moreover, it enables vehicle-to-device communication, vehicle-to-sensor communication, and device-to-device communication within a vehicle. Through the implementing the IoV, smart cities share information on vehicles, roads, infrastructure, buildings, and their surroundings, allowing for the gathering of big data. In addition to providing multimedia and mobile internet application services, the IoV ecosystem offers services for intelligent transportation applications.

In the preliminary data analysis, Fig. 8.2a illustrates the number of documents published per year with the number of publications. The most publications are made in the year 2021 with 98, and as publications are growing each year, this is an excellent field for further research. Articles are the document type with a large number of publications with a percentage of 71.7%. Figure 8.2b shows the document per year by source with subject area. IEEE access has more publications more than 15. Engineering subject domain has most of the publications with a percentage of 33.6%, and Computer Science has the second-most publications with a percentage of 26.9%. Figure 8.2c shows the congregation of keywords that are occurring together in a paper. The interrelations between keywords that are co-appearing in

Fig. 8.1 The Internet of Vehicles (IoV) concept

the research are linked together. Internet of Things, Internet of Vehicles, electric vehicles, and mobile communication are the most co-appearing keywords.

The rest of the chapter is organized as follows. The Internet of Vehicles applications are presented in Sect. 8.2. The connected and autonomous (Level 0 to Level 5) and the layered architecture of the Internet of Vehicle is discussed in Sects. 8.3 and 8.4, respectively. The Internet of Vehicles communication model is presented in Sect. 8.5. The open research challenges and research directions are discussed in Sect. 8.6. Finally, Sect. 8.7 concludes with the summary.

8.2 Internet of Vehicles Applications

Internet of Vehicles (IoV) applications are discussed from two different perspectives in this section:

1. Transportation-related IoV applications
2. Smart city-related IoV applications

Having these two perspectives is justified by the fact that many surveys have focused on the IoV solely serving ITS applications such as traffic efficiency, safety, and infotainment. However, a comprehensive study and investigation of the use of IoV for sensing, collecting, processing, and storing large-scale data has not been

178   8 Internet of Vehicles for Sustainable Smart Cities

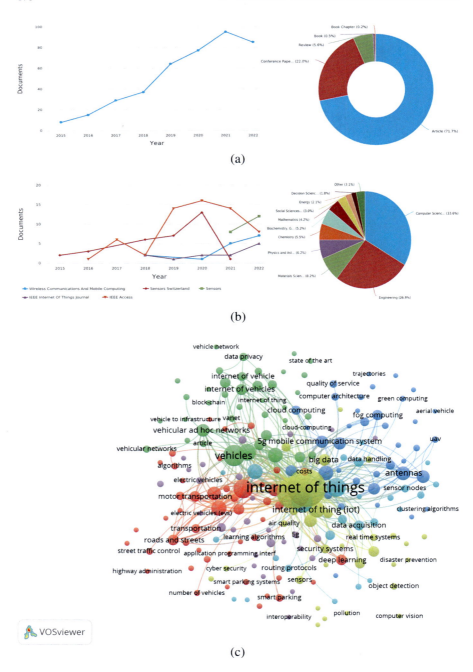

Fig. 8.2 Preliminary data analysis (**a**) documents published per year with the number of publications, (**b**) document per year by source with subject area, (**c**) the congregation of co-appearing keywords from the Scopus database

conducted. While individual vehicle objects and the IoV network itself are the primary focus of ITS-based IoV applications, vehicle objects within smart cities have the capability to provide opportunistic data collection, information transportation, processing, and storage for thousands of objects.

8.2.1 Transportation-Related IoV Applications

The concept of intelligent transport systems (ITS) encompasses a range of technologies and applications that improve travel safety and mobility, increase productivity, and reduce traffic-related problems. The concept of ITS was originally developed by researchers in the United States (US) during the twentieth century [7]. Today, academia and industry are paying close attention to ITS due to its potential to improve vehicle traffic conditions, enhance safety, and contribute to the sustainability of the transportation sector. By preventing traffic congestion and mitigating the effects of climate change on traffic, ITS plays a crucial role. In the transport sector, ITS integrates information and communication technologies (ICTs) [8]. Vehicles and infrastructure are equipped with sensors and equipment that collect data, which is the merged and contextualized to make inferences about a city's transportation system. This data is used to offer services and applications aimed at improving urban resource management and providing information and alert services to enhance convenience for people. By reducing traffic congestion and minimizing the time spent in traffic jams, ITS helps reduce fuel consumption, CO_2 emissions, and monetary losses in cities.

Due to the dynamic and changing nature of vehicle objects in the IoV network, providing vehicles with access to the global internet and maintaining up-to-date information is challenging. The VANET-to-Internet protocol was designed [9] to enable internet access over vehicular networks while ensuring quality of service. In their approach, they utilized the prediction-based routing (PBR) algorithm along with the IEEE 802.11p EDCA scheme. Another study on video streaming and uploading over vehicular networks was proposed [10]. Video streams from moving vehicles are cooperatively forwarded to a fixed network using V2I and V2V communications. According to [11], network coding was proposed as a means to provide streaming over vehicular networks. By recoding received packets, nodes along the transmission path recover lost packets.

8.2.2 Smart City-Related IoV Applications

The concept of intelligent cities, or intelligent spaces as a whole, refers to an environment that is based on the integration of communication and information technologies, as well as sensor systems, into physical objects and the environment in which people live, travel, and work. In order for a city to be intelligent, it must provide solutions to a number of problems, such as traffic, surveillance, natural disasters, and monitoring the environment. Such solutions require the collection and dissemination of urban data through communication infrastructures. In turn, this data flow requires wireless communication that is integrated, heterogeneous, and intelligent. In smart cities, the IoT makes it possible to collect, transport, and process large amounts of data from smart objects at a low cost. In contrast to traditional wireless sensor networks (WSNs), IoV vehicle nodes do not suffer from the limitations of battery power and information processing. Vehicle perform four functions in a smart city system:

- It allows data to be transmitted from one node to another, allowing for more efficient communication and data sharing between different nodes in the network. It also helps to ensure that the network is secure and reliable by maintaining connections between nodes.
- IoV and internet services are consumed by vehicle objects as clients. The nodes in the network are responsible for authenticating each other and validating the data that passes between them. This helps to ensure that the data being sent over the network is coming from a trusted source and accurate, contributing to the maintenance of a secure, reliable, and efficient network for IoV services.
- A vehicle object collects data from other smart objects and transports it to the smart city's data centers. The vehicle object is responsible for encrypting the data and adding a digital signature to it before it is sent over the network. This allows the data center to authenticate the data and verify its integrity. This ensures that the data comes from a trusted source and that it has not been tampered with.
- Vehicle objects provide distributed computing resources to supplement smart objects' constrained information processing capabilities. Vehicle objects also provide additional computing power to the network, allowing for more complex data processing and analysis.

8.3 Connected and Autonomous (Level 0 to Level 5)

In 2025, the researchers predict that there will be approximately 8 million autonomous or semi-autonomous vehicles on the road. This is due to the fact that companies such as Google, BMW, and Tesla are investing heavily in research and development of autonomous vehicle technology. Additionally, many governments are encouraging the adoption of autonomous vehicles by providing incentives such as tax breaks for those who purchase them. Advances in driver assistance technology will see self-driving cars have to pass six levels before merging onto roadways. A scale from 0 (fully manual) to 5 (fully autonomous) is defined by the Society of Automotive Engineers (SAE) as shown in Fig. 8.3. This encourages manufacturers to continue innovating and developing the necessary sensors, algorithms, and other technologies that are needed to make autonomous vehicles a reality. It also helps to ensure that vehicles are safe enough to be on the road before they are released for mainstream use. The US transport department has adopted levels based on these guidelines.

8.3 Connected and Autonomous (Level 0 to Level 5)

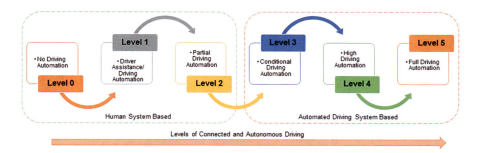

Fig. 8.3 Levels of connected and autonomous (level 0 to level 5)

- *Level 0: No Driving Automation Technology*

Level 0 (zero) automation is the basic level where the driver controls all aspects of the driving task. This means there are no automated systems in the vehicle to monitor the environment or intervene in any way. It is entirely the driver's responsibility to manage the vehicle's movement, including steering, acceleration, braking, parking, and other maneuvers needed to move the vehicle. It is the least automated level of driving, with no automated systems or interventions present. However, at Level 0, there may be driver support systems that temporarily intervene during driving. There are types of stability control, forward-collision warning, automatic emergency braking, blind-spot warning, and lane-keeping assistance. They are considered Level 0 technologies since they do not drive the vehicle, but rather provide alerts or momentary actions in specific circumstances. There are a large number of manual-controlled vehicles on the road today.

- *Level 1 (Driver Assistance/Driving Automation)*

At this level, there is the least amount of automation. There is only one automated system in the vehicle for driver assistance, such as steering or acceleration. With adaptive cruise control, the vehicle stays behind the next vehicle at a safe distance. This is because the human driver monitors other aspects of driving such as steering and braking. Almost all cars equipped with a driver support system have at least one that provides steering, braking, and acceleration assistance at Level 1, the lowest level of automation. At any time and for any reason, the driver must be prepared to take control of the vehicle. Level 1 driver assistance technologies include adaptive cruise control. Drivers do not need to intervene in order to maintain a safe following distance between their vehicles and traffic ahead.

- *Level 2 (Partial Driving Automation)*

An advanced driving assistance system that takes over steering, acceleration, and braking is considered Level 2 driving automation. Although Level 2 driver support control these primary driving tasks, the driver must remain alert and actively supervise the technology at all times. A Level 2 driving automation system is highway driving assist, which is installed in Genesis, Hyundai, and Kia vehicles. On highways, the driver actively steers, accelerates, and brakes the vehicle while holding the steering wheel. Technology from Ford called BlueCruise allows you to travel partially autonomously without having to use your hands. In the United States and Canada, highway driving assist allows drivers to take their hands off the wheel on specific, approved highways. In both of these examples of Level 2 driving automation, the driver must remain alert, engaged, and ready to take control at any moment. When autosteer for city streets comes as an over-the-air software update, it will remain Level 2 technology, as the automaker told California. Vehicles are capable of steering, accelerating, and decelerating. As long as a human sits in the driver's seat, self-driving is not possible since the car can be controlled at any time by the driver. Autopilot by Tesla and Super Cruise by Cadillac (General Motors) are both Level 2 systems.

- *Level 3 (Conditional Driving Automation)*

The third level of automation is known as conditional driving automation. This level of automation allows the car to take control of the steering, acceleration, and braking in certain conditions and provides the driver with an opportunity to take over the driving when needed. Based on changing driving situations around the vehicle, it makes decisions using various driver assistance systems and artificial intelligence (AI). This level of automation is achieved by using sensors, such as cameras and radar, to detect objects in the car's vicinity. The car then uses this data to make decisions and take control of the steering, acceleration, and braking in order to avoid collisions and other dangerous situations. Technology allows people inside the vehicle to engage in other activities without having to supervise them. It is necessary, however, for a human driver to be present and alert and capable of taking control of the vehicle if there is an emergency due to a system failure at any time. Using environmental detection capabilities, Level 3 vehicles make informed decisions about their environment, such as accelerating past slow-moving vehicles. However, they still need to be overridden by humans. When the system fails to perform the task, the driver must remain alert and ready to take control.

- *Level 4 (High Driving Automation)*

A Level 4 vehicle operate autonomously. It is also known as high-driving automation because it does not require any human intervention in the vehicle's operation in the event of a system failure. The lack of a steering wheel and pedals in a Level 4 vehicle is due to the fact that a human driver is never needed. Level 4 vehicles are comfortable enough to let you nap while riding. Driverless taxis and public transportation services will use Level 4 autonomous driving technology. By using geofencing technology, such vehicles will be restricted to specific geographical boundaries from one point to another. During severe weather conditions, Level 4 autonomous vehicles may not be able to operate. A Level 4 vehicle intervene if something goes wrong or there is a system failure, which is what sets it apart from Level 3 automation. In most circumstances, these cars require no human interaction. However, a human is still able to override the system. Nonetheless, this capability is currently limited to a specific area until legislation and infrastructure advancements occur.

- *Level 5 (Full Driving Automation)*

  There is no need for human attention when driving a Level 5 vehicle. It is considered fully autonomous and is capable of navigating any situation or environment with no human input. They are equipped with an array of sensors and cameras that allow them to detect and respond to their surroundings, which eliminates the need for a human driver. The "dynamic driving task" is no longer needed. Steering wheels and acceleration pedals will not even be found in Level 5 cars. Unlike human drivers, they will not be restricted by geofencing and will be able to go anywhere and do anything. The public has yet to experience fully autonomous cars, despite extensive testing around the world.

8.4 Layered Architecture of Internet of Vehicles

The five layered architecture was introduced in [12] and consist of the perception, coordination, AI, application, and business layers. The perception layer is responsible for collecting data from physical sensors and converting them into digital format for further processing. The coordination layer is responsible for managing communication between components in the system. The AI layer is responsible for processing the data and making decisions. The application layer is responsible for controlling the physical components of the system. The business layer is responsible for providing services to users.

- The first layer consists of sensors and actuators integrated into automobiles, remote sensing units, smartphones, and other personal devices. The primary responsibility of this layer is to collect data from vehicles, traffic environments, and connected devices.
- The coordination layer consists of a virtual universal network coordination module that supports heterogeneous networks, including Wi-Fi, 4G/LTE, and WAVE. It ensures that information from the lower layer is transferred to the next AI layer for processing. In this layer, information received from heterogeneous networks is processed according to their different structures. Each candidate network is then analyzed using a unified structure created from the reassembled information.
- As part of the AI layer, the cloud infrastructure also plays a role in third layer. A decision is made about which applications to select based on the information stored, processed, and analyzed in the lower layers.
- The fourth layer represents smart applications such as traffic safety, web-based utility applications, etc. Based on the information processed by the previous layer, this layer aims to provide smart services to end users.
- The fifth layer provides a business model using graphs, flowcharts, tables, and comparisons based on statistical analysis. A heterogeneous network concept is considered in the authors' architecture model.

8.5 Internet of Vehicles Communication Model

Inter-vehicle interaction models perform communication between a vehicle, other vehicles, and the environment. The goal of this section is to provide an overview of current inter-vehicle communication research and current developments in this field. These communication models are Vehicle-to-Vehicle (V2V), Vehicle-to-Infrastructure (V2I), Vehicle-to-Pedestrian (V2P), Vehicle-to-Home (V2H), Vehicle-to-Roadside (V2R), Vehicle to- Everything (V2X), and Vehicle-to-Grid (V2G) models as illustrated in Fig. 8.4.

8.5.1 Vehicle-to-Vehicle (V2V)

The aim of V2V communication is communication among vehicles. This communication enables collaboration and coordination between vehicles, allowing for the dissemination of information without the need for any infrastructure. Due to the unique characteristics of vehicular networks, that is, their high mobility and low communication time, it becomes a challenge to develop communication protocols and efficient data dissemination without impacting the performance of the networks.

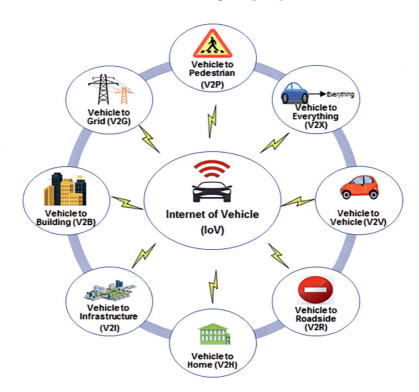

Fig. 8.4 Internet of vehicle communication model

According to [43], V2V communications have two different applications. An accurate positioning system based on vision is presented in the first area to improve vehicle tracking. A framework of several systems was developed to realize a new system of advanced driver assistance systems.

The second area was an attempt to prototype the whole process by combining embedded data, vision data, and virtual-to-virtual simulations in order to progress towards an anti-collision application. An architecture for car tracking and V2V collision warning is proposed that uses real-world GPS data and vision data. The use of matching-based user association methods to optimize information dissemination in IoV aims to maximize information sharing among vehicles [13]. Using peer-to-peer networks, each vehicle establishes multiple and independent connections with other nodes to ensure optimal network connectivity. Using visible light communication (VLC), data is exchanged between vehicles for designated short-range communication and radio frequency [14]. In contrast to other technologies, which share limited bandwidth between a wide variety of applications, VLC enables short-range communication over large, unlicensed, and uncongested bandwidths, which gives it a competitive edge over other technologies.

Using V2V designated short-range communications, an innovative framework is proposed for monitoring and aggregating vehicular traffic information [15]. The performance of dedicated short-range communication in platooning truck scenarios was investigated [16]. Through dedicated short-range communication, vehicles are able to communicate effectively with nearby vehicles and roadside units through wireless radio, which is designed specifically for vehicular environments. It has many benefits, one of which is safety enhancement, which is demonstrated in the study by augmenting the driver's operating process. An intersection collision warning system, for example, emits warning signals using dedicated short-range communication when a vehicle approaches a red traffic light too fast. As a result, other vehicles and pedestrians will be notified in order to prevent collisions.

8.5.2 Vehicle to Roadside (V2R)

Based on random linear network coding, a novel basic safety message (BSM) broadcast scheme is proposed for message dissemination at road intersections [17]. By providing basic safety messages, drivers will gain a better understanding of the local driving environment and avoid collisions at intersections. Despite this, the aforementioned proposal fails to address some of the issues relating to the design of a robust and reliable packet routing protocol for distributing messages at intersections.

To reduce passenger vehicle crashes, V2R communications systems are deployed along roads [18]. By establishing V2R communications between errant vehicles and roadside signs, it may be possible to develop safety systems that detect an oncoming accident and alert the driver. In order to provide flexible and efficient connections between vehicles and roadside units and between vehicle-to-vehicle, ITS services and applications require flexible and efficient connections. Software-defined networking (SDN) has recently prompted the reconsideration of IoV networking paradigms. In addition to improving network flexibility and efficiency, SDN provides a platform for advanced network management.

An SD-IoV framework aims to improve efficiency, flexibility, and QoS assurance through seamless integration of software-defined networking and IoV [19]. It is considered a competitive solution to implement future IoV-based ITS based on SD-IoV, since it inherits the merits of software-defined networking. In [20], a V2R communication data access scheme was proposed that accounted for the integrity of data and its importance. Public transportation vehicles make use of wireless access through V2R communications for streaming purposes. Data streaming over V2R communications links was optimized using a hierarchical optimization framework [21]. It takes into account the needs of the applications, the cost of wireless connectivity for the transit service provider, and the monetary return of the wireless network service operator.

8.5.3 Vehicle to Infrastructure (V2I)

A V2I communication involves vehicle-to-infrastructure communication, which is wireless or cellular. In this communication, internet connectivity is provided in order to enable services on the internet. Safety and efficiency in vehicle infrastructure communications are major targets in V2I communications, according to [22]. Additionally, they explained how vehicular network controllers implement various intra-vehicular communications. A supplementary technology for vehicle-to-vehicle data exchange in vehicular networks is VLC, developed in [14]. The authors suggested that VLC should be implemented together with other wireless standards for future heterogeneous vehicular networks based on some of the identified limitations of pure vehicular visible light networks (VVLNs). The Social Internet of Vehicles (SIoV), which is a vehicular case of the Social Internet of Things (SIoT), has been proposed [23]. Through the SIoV, existing technologies for V2I communications will be leveraged. In their investigation of Mobile Content Delivery Vehicular Networks (MCDVNs), [24] proposed the sigma deployment strategy for measuring content delivery performance in infrastructure-based vehicular networks. According to [25], resource allocation in a multi-rate multiple lane V2I network for drive-thru internet applications are based on proportional fairness. Several vehicles transmit at different data rates and move at different mean velocities on the same RSU. Taking into account the different data update rates for road information and the present evolution of V2I standards proposed a novel auxiliary ZigBee-based architecture for the AUTOPIA communications scheme, which will enable vehicles to obtain road information from the infrastructure and adapt themselves to the current zone configuration without the need for implementing complex routing maps [26]. They proposed designing and implementing a low-cost infrastructure network based on ZigBee technology to alert drivers to unexpected circumstances on the road, enabling them to make appropriate decisions.

In [27], various network technologies in the field of vehicle communications were examined, including Bluetooth, ZigBee, Ultra-Wide Band (UWB), Wi-Fi, etc., and the use of cellular networks (CN) for both V2V and V2I communication was recommended. An investigation of heterogeneous wireless networks based on Wi-Fi, DSRC, and LTE technologies was conducted [28]. According to [29], traffic signals are one of the major reasons for travel delays in urban settings, despite their importance for safe traffic operations at busy intersections. They proposed an alternative to traffic signals by using Advisory Speed Limit (ASL) control strategies based on V2I communication. In [30], coverage and capacity requirements are analyzed for the implementation of V2I communications using digital broadcasting, cellular communication, and dedicated short-range communication systems. Digital broadcasting systems are shown to have inherent capacity limitations. To implement V2I links, Universal Mobile Telecommunications Systems (UMTS) are expected to use a dedicated channel (DCH) or a multimedia broadcast/multicast service (MBMS). Connecting distant nodes in a wireless network is typically accomplished through multi-hop communication. Using cooperative vehicles serving as intermediate relays, an isolated vehicle establishes multi-hop connectivity with a faraway roadside unit [31].

8.5.4 Vehicle to Home (V2H)

A V2H operating system provides backup power in the event of a power outage on the external electric grid, according to Shin and Baldick [32]. It proposes a new algorithm for V2H systems and introduces a novel optimization model aimed at maximizing backup duration. In their study, V2Hs were considered as an extended system for supplying backup power to homes without energy resources or for providing more balanced backup durations.

8.5.5 Vehicle to Everything (V2X)

One of the most challenging aspects of V2X (including V2V and V2I) communications is designing the user association strategy, which determines how vehicles should communicate with other vehicles and Roadside Units (RSUs) to increase the diversity of data transmitted through the network. In order to achieve this, authors in [13] examined the user association problem in IoV, aiming to optimize the dissemination of information within the network. They proposed an independent user association method, modeled after Irving's Stable Fixture (ISF), where each vehicle establishes independent links with other vehicles/infrastructures without requiring its partners to be connected to each other. The use cases and service requirements for Long-Term Evolution (LTE)-based V2X were presented in [33]. Moreover, they examined the major technical challenges for LTE-based V2X and proposed solutions. A V2V communication is classified as a V2V technology, while V2I, V2N, and V2P are classified as a V2X technology. As a systematic and integrated solution for V2X communications, the authors considered LTE-based V2X, motivated both by technical and industrial perspectives. They noted that with an LTE network, ubiquitous coverage is achieved, and interoperability with commercial operators is possible. Using two configurations of a two-element monopole array and a switchless reconfigurable antenna, in [34] a new concept for rooftop mounted vehicular multiple-input multiple-output (MIMO) antennas for V2X communications was investigated, determining the best suited antenna type for each use case. In this scenario, buildings at a blind corner intersection block line-of-sight (LOS) between two vehicles, and interference avoidance is the goal. According to [35], they outlined the service flow and requirements for V2X services that LTE systems target and discussed scenarios suitable for operating LTE-based V2X services. They addressed major challenges, such as high mobility and densely populated vehicle environments, to fulfil V2X service requirements.

8.5.6 Vehicle to Grid (V2G)

The use of V2G systems brings many benefits to power systems, such as stabilizing energy demand and supply fluctuations and assisting PEV users in reducing energy costs [36]. A 6LowPAN wireless sensor network and the constrained application protocol (CoAP) are used in the open implementation of (automatic dependent rebroadcast) ADR [37]. An example of CoAP's reliability and suitability its use in the charging of plug-in hybrid electric vehicle (PHEV), which requires a self-service kiosk that accepts credit cards and controls the charging process. In V2G, [38] proposed the use of the LTE protocol to facilitate communication between electric vehicles. For the development of V2G communication in smart grids, this research serves as a platform for 4G/5G networks. A mathematical model and practical implementation of an innovative V2G system with networked electric vehicles (NEVs) was proposed in [39]. Using partial differential equation (PDE)-based techniques, in [40], model and control a large population of grid connected plug-in electric vehicles (PEVs). According to [41], office buildings equipped with EV charging stations employ customizable V2G smart charging coordination systems. Through V2G coordination, the proposed scheme reduces energy costs by optimizing the building's energy consumption and electricity load. In the context of smart buildings, the analysis centers around coordinating the charging and discharging of electric vehicles. The efficiency of V2G systems is inevitably affected by the vulnerabilities in V2G communication and the cyber security threats they encounter. Consequently, a distinct approach and focus are required to mitigate the cyber risks associated with V2G systems.

8.5.7 Vehicle to Pedestrian (V2P)

According to [42], two novel methods are proposed to address the issue of collective scanning and extension receiving in V2P communication, which utilizes a wireless local area network (WLAN). Despite the increased safety and convenience provided by highly computerized vehicles', the number of fatal accidents involving pedestrians at night has risen in recent years. To mitigate this, an illumination system utilizing the LED projection module is proposed in [43]. It aims to reduce pedestrian accidents by projecting images on the road surface to facilitate communication between cars and pedestrians. For vulnerable road users, a cooperative system combining perception with peer-to-peer communication is proposed [44].

8.6 Open Research Challenges and Research Directions

In the IoV ecosystem, one of the main challenges is the integration of all components and object communications. To meet these requirements, several authors have proposed IoV architectures based on layered designs [45–49]. However, it is important to note that these designs are specifically tailored for ITS applications and do not consider the general scenario of smart city environments or big data processing. Data security is crucial in the IoV due to the integration of various technologies, services, and standards, as open and public networks are vulnerable to intrusions and cyber-attacks that cause physical damage and privacy breaches.

Another challenge is the rapid growth of the IoV and the massive amount of data is collected between vehicles and application platforms. This challenge corresponds to the heterogeneous nature of big data in terms of size, volume, and dimensionality. Therefore, big data processing and analytics are necessary for both research and commercial purposes. However, many existing proposed IoV architectures [50–53], inadequately consider or completely overlook big data collection, processing, and analytics. Recently, [54] focused solely on the secure mechanism for IoV big data collection. Furthermore, the increasing proliferation of multimedia content within vehicles and between vehicles and applications generates an enormous amount of multimedia data, including text, audio, images, and video. Thus, there is a new universal IoV architecture that address these challenges effectively. The deployment of the IoV in smart cities poses technological and social challenges. The increasing number of connected vehicles in smart cities implies more vehicle nodes (moving nodes) before the collection of big data begins. This has implications on sink nodes after packets are successfully delivered and the packet delivery rate depends on the speed of vehicle nodes. As the IoV network structure evolves, the packet delivery success rate becomes more complex and existing IoT works do not account for these factors, making direct application to IoV with big data challenging. Following are the points related to challenges and future directions:

- Security needs to be addressed from a social perspective, such as through policy development, law enforcement, and security awareness training.
- Research needs to investigate how human agents interact with technical security solutions. There is a lot of research on social aspects of social network security, which can be borrowed for this purpose.
- It is important to examine the social perspective of security and the interaction between human agents and technical solutions in future research.
- Both social networks and vehicular networks can benefit from future privacy-enhancing technology in IoV.
- To thoroughly evaluate the upcoming IoV privacy technologies, privacy metrics from both domains must be considered.

An increasingly critical challenge in intelligent cities is the analysis and interpretation of data obtained from mobile and fixed devices with localization and communication capabilities, such as GPS, Wi-Fi, 3G, and 4G. To maximize connectivity and forecasting, individuals' mobility must be taken into account in such environments. Understand the dynamic nature of the urban environment, it is necessary to identify patterns of mobility, social events, routines, and the interactions between individuals and their surroundings [55]. Microscopic factors linked to human behavior, when coupled with urban geographical characteristics, allow identification of critical regions in a street network based on trends and flow aspects of individuals over time. This means that deep analysis identifies the most frequented regions at a given time and why they receive a given stream of people. Designing a robust, reliable solution for real-time data collection in smart cities is one of the most challenging tasks. In addition, the precision of such a solution in monitoring and interpreting data for decision-making is another challenge. It should be possible to use the analysis output to allocate or move physical resources not only for citizens but also for municipal services, such as fire departments, ambulances, and police.

8.7 Summary

The Internet of Vehicles is a revolutionary type of network that enables social interactions among vehicles, drivers, and passengers. The IoV is a special application within the IoT sector. It has gained much attention and has become an essential platform for interacting between vehicles, humans, and roadside infrastructures. Overtime, IoTs will become an invaluable part of our lives, allowing us to avoid traffic lights, road accidents, and other transportation-related issues. The updated system will make traffic services more convenient, comfortable, and safe for millions of people.

It is evolving into a global heterogeneous network of vehicles. In addition to automating the security and efficiency features of vehicles, IoV also aims to commercialize vehicular networks. Understanding the layered architecture, network model, and challenges of IoV would be beneficial. Developers will benefit from technology-oriented application development and technological advancements based on applications. It is imperative for researchers to examine the identified challenges and issues in the design and development of IoV.

References

1. A. Caragliu, C. Del Bo, P. Nijkamp, Smart cities in Europe. J. Urban Technol. **18**, 65–82 (2011)
2. T. Ishida, K. Isbister, *Digital Cities: Technologies, Experiences, and Future Perspectives* (Springer, Berlin, 2000)
3. Z. Xiong, H. Sheng, W. Rong, D.E. Cooper, Intelligent transportation systems for smart cities: A progress review. SCIENCE CHINA Inf. Sci. **55**, 2908–2914 (2012)

References

4. R. Khatoun, S. Zeadally, Smart cities: Concepts, architectures, research opportunities. Commun. ACM **59**(8), 46–57 (2016)
5. K. Su, J. Li, H. Fu, Smart city and the applications, in *Proceedings of the 2011 International Conference on Electronics, Communications and Control (ICECC)*, (Zhejiang, China, 9–11 September 2011), pp. 1028–1031
6. R.G. Hollands, Will the real smart city please stand up? Intelligent, progressive or entrepreneurial? City **12**(3), 303–320 (2008)
7. K. Jadaan, S. Zeater, Y. Abukhalil, Connected vehicles: An innovative transport technology. Procedia Eng. **187**, 641–648 (2017)
8. European Commission, European commission mandate m/453en (2019). Available online: https://law.resource.org/pub/eu/mandates/m453.pdf. Accessed on 23 Jan 2023
9. A. Ksentini, H. Tounsi, M. Frikha, A proxy-based framework for QoS-enabled Internet access in VANETS, in *Proceedings of the 2nd International Conference on Networking, Communications*, (November 2010), pp. 1–4
10. X. Zhu, J. Bi, M. Li, H. Liu, Adaptive video streaming uploading with moving prediction in VANETs scenarios, in *Proceedings of IEEE International Conference on Networking, Architecture, and Storage*, (August 2015), pp. 39–44
11. A. Razzaq, A. Mehaoua, Video transport over VANETs: Multistream coding with multipath and network coding, in *Proceedings of IEEE Conference on Local Computer Networks*, (October 2010), pp. 32–39
12. O. Kaiwartya et al., Internet of vehicles: Motivation, layered architecture, network model, challenges, and future aspects. IEEE Access **4**, 5356–5373 (2016)
13. F. Chiti, R. Fantacci, Y. Gu, Z. Han, Content sharing in internet of vehicles: Two matching-based user-association approaches. Veh. Commun. **8**, 35–44 (2017)
14. A. Bazzi, B.M. Masini, A. Zanella, A. Calisti, Visible light communications as a complementary technology for the internet of vehicles. Comput. Commun. **93**, 39–51 (2016)
15. Y. Lou, P. Li, X. Hong, A distributed framework for networkwide traffic monitoring and platoon information aggregation using V2Vcommunications. Transp. Res. C, Emerg. Technol. **69**, 356–374 (2016)
16. S. Gao, A. Lim, D. Bevly, An empirical study of DSRC V2V performance in truck platooning scenarios. Digit. Commun. Netw. **2**(4), 233–244 (2016)
17. Y. Gao, G.G.M.N. Ali, P.H.J. Chong, Y.L. Guan, Network coding based BSM broadcasting at road intersection in V2V communication, in *Proceedings of IEEE Vehicular Technology Conference (VTC-Fall)*, (September 2016), pp. 1–5
18. S. Temel, M.C. Vuran, R.K. Faller, A primer on vehicle-to-barrier communications: Effects of roadside barriers, encroachment, and vehicle braking, in *Proceedings of Vehicular Technology Conference*, (September 2016), pp. 1–7
19. J. Chen et al., Service-oriented dynamic connection management for software-defined internet of vehicles. IEEE Trans. Intell. Transp. Syst. **18**(10), 2826–2837 (2017)
20. T. Guo, C. Li, Y. Zhang, Z. Miao, L. Xiong, Integrity-oriented service scheduling for vehicle-to-roadside data access, in *Proceedings of 19th International Symposium on Wireless Personal Multimedia Communications (WPMC)*, (November 2016), pp. 377–382
21. D. Niyato, E. Hossain, A unified framework for optimal wireless access for data streaming over vehicle-to-roadside communications. IEEE Trans. Veh. Technol. **59**(6), 3025–3035 (2010)
22. S. Habib, M.A. Hannan, M.S. Javadi, S.A. Samad, A.M. Muad, A. Hussain, Inter-vehicle wireless communications technologies, issues and challenges. Inf. Technol. J. **12**(4), 558–568 (2013)
23. K.M. Alam, M. Saini, A.E. Saddik, Toward social internet of vehicles: Concept, architecture, and applications. IEEE Access **3**, 343–357 (2015)
24. C.M. Silva, F.A. Silva, J.F.M. Sarubbi, T.R. Oliveira, W. Meira Jr., J.M.S. Nogueira, Designing mobile content delivery networks for the internet of vehicles. Veh. Commun. **8**, 45–55 (2017)
25. V.P. Harigovindan, A.V. Babu, L. Jacob, Proportional fair resource allocation in vehicle-to-infrastructure networks for drive-thru internet applications. Comput. Commun. **40**, 33–50 (2014)

26. J. Godoy, V. Milanes, J. Perez, J. Villagra, E. Onieva, An auxiliary V2I network for road transport and dynamic environments. Transp. Res. C, Emerg. Technol. **37**, 145–156 (2013)
27. J. Santa, A.F. Gomez-Skarmeta, M. Sanchez-Artigas, Architecture and evaluation of a uni_ed V2V and V2I communication system based on cellular networks. Comput. Commun. **31**, 2850–2861 (2008)
28. K.C. Dey, A. Rayamajhi, M. Chowdhury, P. Bhavsar, J. Martin, Vehicle-to-vehicle (V2V) and vehicle-to-infrastructure (V2I) communication in a heterogeneous wireless network—Performance evaluation. Transp. Res. C, Emerg. Technol. **68**, 168–184 (2016)
29. G.A. Ubiergo, W.L. Jin, Mobility and environment improvement of signalized networks through vehicle-to-infrastructure (V2I) communications. Transp. Res. C, Emerg. Technol. **68**, 70–82 (2016)
30. P. Belanovic, D. Valerio, A. Paier, T. Zemen, F. Ricciato, C.F. Mecklenbrauker, On wireless links for vehicle-to-infrastructure communications. IEEE Trans. Veh. Technol. **59**(1), 269–282 (2010)
31. R. Atallah, M. Khabbaz, C. Assi, Multihop V2I communications: A feasibility study, modeling, and performance analysis. IEEE Trans. Veh. Technol. **66**(3), 2801–2810 (2017)
32. H. Shin, R. Baldick, Plug-in electric vehicle to home (V2H) operation under a grid outage. IEEE Trans. Smart Grid **8**(4), 2032–2041 (2017)
33. K. Sung, J. Lee, J. Shin, Study of CAN-to-3GPP LTE gateway architecture for automotive safety in V2I environment, in *Proceedings of 17th International Conference on Advanced Communication Technology (ICACT)*, (IEEE, 2015), pp. 256–259
34. T. Kopacz, A. Narbudowicz, D. Heberling, M. J. Ammann, Evaluation of automotive MIMO antennas for V2V communication in urban intersection scenarios, in *Proceedings of 11th European Conference on Antennas and Propagation (EUCAP)*, (March 2017), pp. 2907–2911
35. H. Seo, K.-D. Lee, S. Yasukawa, Y. Peng, P. Sartori, LTE evolution for vehicle-to-everything services. IEEE Commun. Mag. **54**(6), 22–28 (2016)
36. D. Niyato, D.T. Hoang, P. Wang, Z. Han, Cyber insurance for plug-in electric vehicle charging in vehicle-to-grid systems. IEEE Netw. **31**(2), 38–46 (2017)
37. A. Fechechi et al., A new vehicle-to-grid system for battery charging exploiting IoT protocols, in *Proceedings IEEE International Conference on Industrial Technology*, (March 2015), pp. 2154–2159
38. S. Kumar, R.Y.U. Kumar, Performance analysis of LTE protocol for EV to EV communication in vehicle-to-grid (V2G), in *Proceedings of IEEE 28th Canadian Conference on Electrical and Computer Engineering (CCECE)*, (May 2015), pp. 1567–1571
39. R. Huang et al., Integration of IEC 61850 into a vehicle-to-grid system with networked electric vehicles, in *Proceedings of Power & Energy Society Innovative Smart Grid Technologies Conference (ISGT)*, (IEEE, 2015), pp. 1–5
40. C. Le Floch, E.C. Kara, S. Moura, PDE modeling and control of electric vehicle fleets for ancillary services: A discrete charging case. IEEE Trans. Smart Grid **9**(2), 573–581 (2016)
41. M. Shurrab, S. Singh, H. Otrok, R. Mizouni, V. Khadkikar, H. Zeineldin, An efficient vehicle-to-vehicle (V2V) energy sharing framework. IEEE Internet Things J. **9**(7), 5315–5328 (2021)
42. S. Fujikami, T. Sumi, R. Yagiu, Y. Nagai, Fast device discovery for vehicle-to-pedestrian communication using wireless LAN, in *Proceedings of 12th Annual IEEE Consumer Communications and Networking Conference (CCNC)*, (January 2015), pp. 35–40
43. M. Suwa, M. Nishimuta, R. Sakata, LED projection module enables a vehicle to communicate with pedestrians and other vehicles, in *Proceedings of IEEE International Conference on Consumer Electronics*, (January 2017), pp. 37–38
44. P. Merdrignac, O. Shagdar, F. Nashashibi, Fusion of perception and V2P communication systems for the safety of vulnerable road users. IEEE Trans. Intell. Transp. Syst. **18**(7), 1740–1751 (2017)
45. N. Liu, Internet of vehicles your next connection. Huawei WinWin **11**, 23–28 (2011)
46. K. Golestan, R. Soua, F. Karray, M. Kamel, Situation awareness within the context of connected cars: A comprehensive review and recent trends. Inf. Fusion **29**, 68–83 (2016)

References

47. J. Wan, D. Zhang, S. Zhao, L.T. Yang, J. Lloret, Context-aware vehicular cyber-physical systems with cloud support: Architecture, challenges, and solutions. IEEE Commun. Mag. **52**(8), 106–113 (2014)
48. P. Gandotra, R.K. Jha, S. Jain, A survey on device-to-device (D2D) communication: Architecture and security issues. J. Netw. Comput. Appl. **78**, 9–29 (2017)
49. F. Bonomi, *The Smart and Connected Vehicle and the Internet of Things* (2013). [Online]. Available http://tf.nist.gov/seminars/WSTS/PDFs/1-0_Cisco_FBonomi_ConnectedVehicles.pdf. Accessed on 8 Mar 2023
50. B. Ji, X. Zhang, S. Mumtaz, C. Han, C. Li, H. Wen, D. Wang, Survey on the internet of vehicles: Network architectures and applications. IEEE Commun. Stand. Mag. **4**(1), 34–41 (2020)
51. F. Yang, J. Li, T. Lei, S. Wang, Architecture and key technologies for internet of vehicles: A survey. J. Commun. Inf. Netw. **2**(2), 1–17 (2017)
52. J. Contreras-Castillo, S. Zeadally, J.A. Guerrero-Ibanez, A seven layered model architecture for internet of vehicles. J. Inf. Telecommun. **1**(1), 4–22 (2017)
53. J. Contreras-Castillo, S. Zeadally, J.A. Guerrero-Ibanez, Internet of vehicles: Architecture, protocols, and security. IEEE Internet Things J. **5**(5), 3701–3709 (2018)
54. L. Guo et al., A secure mechanism for big data collection in large scale internet of vehicle. IEEE Internet Things J. **4**(2), 601–610 (2017)
55. G. Pan, G. Qi, W. Zhang, S. Li, Z. Wu, L. Yang, Trace analysis and mining for smart cities: Issues, methods, and applications. IEEE Commun. Mag. **51**(6), 120–126 (2013)

Chapter 9
Smart Healthcare in Sustainable Smart Cities

9.1 Introduction

Due to the enormous growth in population, traditional healthcare is unable to meet the needs of everyone. It is not affordable or accessible for everyone to access medical services, despite excellent infrastructure and cutting-edge technologies. Smart healthcare aims to educate and keep users aware of their health status by educating them about their medical status. Smart healthcare emerges as a response to the challenges faced by traditional healthcare systems, which are constrained by limited resources and a growing demand for services. It aims to make healthcare more intelligent, efficient, and sustainable by leveraging advanced technologies and innovative approaches, thereby enabling self-management of certain emergencies [1]. Using available resources efficiently is the key to smart healthcare. The technology facilitates remote monitoring of patients and reduces the cost of treatment for users. In addition, it facilitates the expansion of medical practitioners' services across regional boundaries. A smart healthcare system ensures that its citizens lead healthy lives in a smart city, which aligns with the growing trend toward smart cities.

Smart healthcare in sustainable smart cities is a concept that combines the use of technology, data, and analytics to improve the delivery of healthcare in cities. It seeks to improve the health and well-being of citizens while also making cities more sustainable and efficient. It has the potential to revolutionize urban living, allowing for more sustainable smart cities. By providing better access to healthcare, smart healthcare improve the overall health of citizens, reducing the burden on existing healthcare infrastructure and allowing cities to become smarter and more sustainable. Smart healthcare brings with it the potential to reduce healthcare costs by providing more accurate diagnoses and treatments, thus minimizing the need for costly hospital visits. Smart healthcare is the use of technology to improve the delivery of healthcare services. It includes the use of electronic health records, telemedicine, and other health information technology. This ecosystem uses sensors to perceive data, transmits data via the Internet of Things (IoT), and processes data using cloud computing [2]. A dynamic and refined management of human society is realized through the coordination and integration of social systems, enabling smart healthcare to enhance the delivery of healthcare services. Smart healthcare aims to provide individuals with more personalized and effective healthcare services, reduce costs, and improve the efficiency of healthcare delivery, ultimately enhancing the quality of life for many through better access to medical information and services.

The concept of smart healthcare is defined as an advanced stage of medical technology development enabling dynamic access to information through wearable devices, the IoT, and the m-Internet (mobile internet). It connects individuals, materials, and institutions associated with healthcare, actively manages and responds to the needs of the medical ecosystem, promotes interaction between all parties involved in the healthcare field, helps them make informed decisions, and facilitate the rational allocation of resources [3].

The term "connected health" refers to any digital healthcare solution capable of remote monitoring, including constant health monitoring, emergency response, and alarm functionality. The aims of connected health are to improve healthcare quality and efficiency by enabling self-care and complementing it with remote care. The concept comes from telemedicine, when users receive feedback about their health whenever needed. A smart healthcare solution operates completely autonomously, whereas a connected healthcare approach provides clinical feedback to users. Smart healthcare is classified according to how it is consumed by end users, which determines the structure and dynamics of the industry.

Based on the services, medical devices, technologies used, applications, system management, and end users, Fig. 9.1 illustrates the broad classification of the smart healthcare sector. The use of connectivity technologies expands the applications for which the healthcare system is designed. By integrating small devices through wireless technologies, remote health monitoring is implemented via IoT [4]. It is, however, necessary to maintain constant internet connectivity and support heavy data traffic in a hospital setting where a healthcare network is managed using Wi-Fi and cables.

Digital healthcare solutions and systems that are fully connected and accessible remotely are called smart and connected health [5]. It was launched in 2013 by the National Science Foundation (NSF) and the National Institutes of Health (NIH) in order to accelerate the development and integration of innovative information technology approaches to health [6]. The research team envisioned developing innovative "smart" ideas through multidisciplinary collaboration in order to enhance existing and new scientific collaborations. Healthcare delivery has benefited from technology, especially artificial intelligence (AI), under the umbrella of smart cities. Medical informatics, public health, big data, bio-engineering, the telecommunications industry, and many other disciplines are involved in smart and connected health research. In different critical health contexts, smart and connected health may require resource-aware, time-constrained, complex, and secure healthcare transactions among multiple stockholders. In addition to revolutionizing next-generation

9.1 Introduction

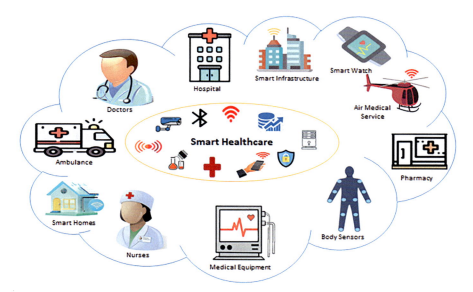

Fig. 9.1 Smart and connected healthcare system

healthcare, smart and connected health has an array of significant benefits, including accelerating treatment and testing procedures, reducing physician visits, responding efficiently to emergencies and pandemic outbreaks, and improving the quality of patient care [7].

In preliminary data analysis, Fig. 9.2a illustrates the number of documents published per year with the number of publications. The most publications are made in the year 2022 with 523, and as publications are growing each year, this is an excellent field for further research. Articles are the document type with a large number of publications with a percentage of 66.2%. Figure 9.2b shows the document per year by source with subject area. Sensors has more publications more than 60. Computer Science subject domain has most of the publications with a percentage of 30.8%, and Engineering has the second most publications with a percentage of 24.4%. Figure 9.2c shows the congregation of keywords that are occurring together in a paper. The interrelations between keywords that are co-appearing in the research are linked together. Internet of Things, healthcare, blockchain, big data, wearable sensors, and smartphones are the most co-appearing keywords.

The rest of the chapter is organized as follows. Healthcare 4.0 is explained in Sect. 9.2. The Internet of Medical Things (IoMT) and smart healthcare services are presented in Sects. 9.3 and 9.4, respectively. Various enabling technologies and applications for sustainable smart cities are discussed in Sects. 9.5 and 9.6, respectively. The open research challenges and research directions are discussed in Sect. 9.7. Finally, Sect. 9.8 concludes with the summary.

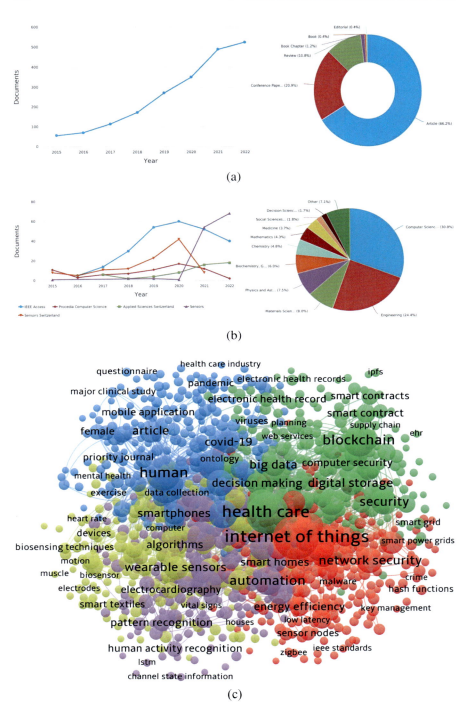

Fig. 9.2 Preliminary data analysis (**a**) the published documents per year with the number of publications, (**b**) source with subject area, and (**c**) the congregation of co-appearing keywords from the Scopus database

9.2 Healthcare 4.0

Healthcare 4.0 is the application of modern technologies such as AI, machine learning (ML), robotics, and the IoT to healthcare. It allows for better, faster, and more personalized healthcare, including the use of predictive analytics to diagnose and treat diseases and other medical conditions. It promises to revolutionize healthcare by enabling more accurate and faster diagnoses, treatments tailored to individual patients, and a reduction in healthcare costs. This is possible because with predictive analytics, healthcare providers have access to vast amounts of patient data that is used to identify patterns and trends in illnesses and treatments. This data then be used to create more accurate and personalized treatment plans, as well as to identify potential patients at risk of certain illnesses. Additionally, predictive analytics helps to reduce healthcare costs by predicting and preventing potential problems before they occur. Healthcare delivery has experienced a long history of evolutions and revolutions. There are multiple stages represent the evolution from Healthcare 1.0 to Healthcare 4.0.

- **Healthcare 1.0**

  It refers to the basic encounter between a patient and a physician. Patients visit clinics and meet with physicians and other healthcare practitioners during such visits. Clinicians perform consultations, tests, diagnoses, and prescribe medications, as well as follow-up plans (e.g., ordering lab tests or imaging tests, referring patients to specialists). This model has been used in healthcare for hundreds of years.

- **Healthcare 2.0**

  The concept of Healthcare 2.0 emphasizes the use of technology and data to improve patient care. Rather than focussing solely on treating symptoms, Healthcare 2.0 takes a holistic approach to patientcare, such as Magnetic Resonance Imaging (MRI), ultrasound, Computed Tomography (CT) scanning, pulse oximeter, and arterial lines. A variety of technologies are utilized, including electronic health records, telemedicine, and big data analytics. To provide personalized care, Healthcare 2.0 collects, stores, and analyzes data from patients using digital tools. Besides monitoring patients remotely, it provide healthcare providers with real-time feedback. Moreover, Healthcare 2.0 facilitates collaboration among healthcare providers, leading to improved outcomes.

- **Healthcare 3.0**

  In addition to being more efficient and cost-effective, this type of healthcare delivery offers increased convenience for both patients and providers. Furthermore, it facilitated improved communication between patients and providers, providing enhanced access to care, particularly for individuals residing in rural areas. Moreover, it enables for more accurate diagnoses, effective treatment, and ultimately, improved patient outcomes. The availability of computer networks has made remote care and telehealth possible, and electronic visits (such as communicating with a physician through a patient portal) are beginning to replace some face-to-face encounters. Due to ongoing COVID-19 pandemic, telehealth and virtual visits have experienced increased popularity.

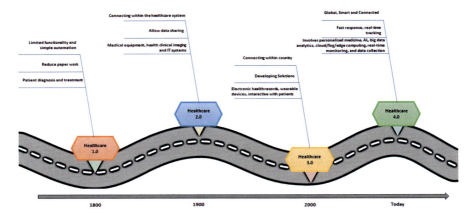

Fig. 9.3 Healthcare 1.0 to 4.0

- **Healthcare 4.0**

In Healthcare 4.0, the healthcare delivery process becomes a cyber physical system (CPS) equipped with IoT, RFID (radio frequency identification), wearables, and various medical devices, intelligent sensors, medical robots, and more. With the technologies, such as cloud computing, big data analysis, AI, and decision support, healthcare is delivered smartly and interconnectedly. It connects not only healthcare organizations and facilities (e.g., hospitals, clinics, and long-term care facilities) but also all the equipment and devices, as well as patients' homes and communities. We also envision proactive treatments, disease prediction and prevention, personalized medicine, and enhanced patient-centered care through AI techniques. These advancement leads to the paradigm of Healthcare 4.0, which represents a pervasive, intelligent, and interconnected healthcare community. The transitions from Healthcare 1.0 to 4.0 are depicted in Fig. 9.3.

9.3 The Internet of Medical Things

The Internet of Medical Things (IoMT) is comprised of medical equipment and applications connected to the internet. This technology allows medical professionals to collect and analyze data from medical devices in real time, enabling them to make more informed decisions. IoMT also enables remote patient monitoring, which reduce healthcare costs and improve outcomes. There are numerous use cases for IoMT, benefiting patients, medical professionals, researchers, and insurers. These include immediate medical help, data analysis, prescription management, augmented operations, monitoring patient and employee, inventory management, and more, as illustrated in Fig. 9.4. By 2025, it is estimated that IoT devices will be utilized in healthcare, accounting for approximately 40% of the total value of IoT technology [8]. Additionally, by the end of 2019, about 87% of healthcare organizations were expected to have adopted IoT solutions [9].

9.4 Smart Healthcare Services

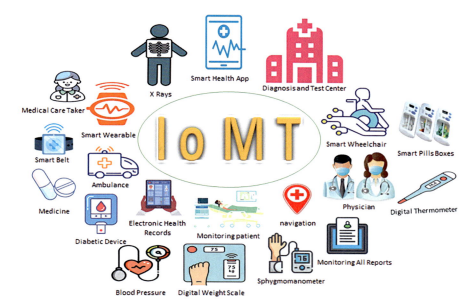

Fig. 9.4 Concept of Internet of Medical Things (IoMT)

Healthcare services based on mobile computing, medical sensors, and communications technologies are known as m-health. It is an emerging field that offers improved access to healthcare services for both patients and providers, providing an efficient and cost-effective way to deliver healthcare services remotely. This includes telemedicine, telemonitoring, and patient education. With m-health, healthcare professionals remotely monitors patients, diagnose and treat them, and provide information and support. Furthermore, m-health has the potential to enhance the quality of care, reduce healthcare costs, and increase accessibility, particularly for underserved populations. By integrating 6LowPAN (IPv6 over Low-Power Wireless Personal Area Network) with evolving 4G networks, m-IoT enables the development of future internet-based m-health services. While m-IoT refers to the IoT for healthcare services, it is important to note that it incorporates characteristics of global participant mobility. According to [10], m-IoT services have been investigated for noninvasively glucose level measurements through the use of m-IoT architecture and implementation issues. However, the low network power consumption of a message-exchange-based mobility system has not been verified in [11].

9.4 Smart Healthcare Services

Smart healthcare services use modern technology like AI, ML, and IoT to automate tasks, collect patient data, and improve the accuracy and effectiveness of patient care. This helps to reduce medical errors and improve patient outcomes. This makes healthcare services more efficient, cost-effective, and reliable, providing patients with better care and quality of life. Some of the services are discussed as follows:

- **Healthcare Monitoring Devices**

IoT devices provide healthcare professionals and patients with a number of new monitoring opportunities. These devices track a patient's vital signs and other data in real time, allowing healthcare professionals to monitor a patient's health remotely and intervene if necessary. This improve patient outcomes and reduce the need for hospital visits. Healthcare providers and their patients alike benefit from wearable IoT devices and face a number of challenges as well. Wearable devices like those described above remain the most common type of IoT device in healthcare, but there are also devices that provide treatment, or even reside within or on the patient.

- **Remote Patient Monitoring**

IoT devices are most commonly used for remote patient monitoring in healthcare. This technology allows patients to be monitored in real time, even when they are outside of a healthcare facility. IoT devices are able to track vital signs such as heart rate, blood pressure, and temperature and also monitors medication intake and activity levels. This allows healthcare professionals to assess a patient's health quickly and accurately, providing timely and appropriate care. Health metrics like heart rate, blood pressure, temperature, and more are automatically collected by IoT devices from patients who are not physically present in healthcare facilities, eliminating the need for patients to travel to their providers or collect the data themselves. The collected data is transmitted to a software application accessed by healthcare professionals and/or patients. Algorithms are then used to analyze the data, recommend treatments and generate alerts. While remote patient monitoring has the potential to improve patient outcomes, there are privacy and security concerns that need to be addressed. The major challenge is the risk of hackers gaining unauthorized access to patient data transmitted wirelessly. Additionally, there is a concern that insurance companies could use patient data to discriminate against individuals with pre-existing conditions.

- **Glucose Monitoring in Healthcare**

Monitoring glucose levels is important for people with diabetes, to help them manage their blood sugar levels. It is used to monitor other conditions, such as hypoglycemia, which is caused by certain medications or illnesses. By monitoring glucose levels closely, individuals with diabetes are proactive in managing their health and avoiding the risk of developing potentially serious complications. Measuring and recording glucose levels manually, takes a great deal time, and this is not only inconvenient, it also only reports a patient's glucose level at the moment of the test. In cases where levels fluctuate widely, periodic testing may not be sufficient to detect problems. By providing continuous, automatic glucose monitoring to patients, IoT devices helps to address these challenges. Patients are alerted when their glucose levels are abnormal when glucose monitoring devices are used instead of manual record keeping.

For instance, if a person notices that their glucose levels are starting to decline, they should take steps to increase their glucose levels before they become dangerously low. However, there are also potential risks associated with always on glucose monitoring. For example, if data from the device is not transmitted properly, or is not interpreted correctly, it could lead to false alarms or inaccurate readings. This

9.4 Smart Healthcare Services

could cause unnecessary anxiety for patients or, worse, lead to dangerous situations if patients do not take appropriate actions in response to the alarms.

- **Heart Rate Monitoring**

  Heart rate monitoring helps to measure the intensity of physical activity and determine the effectiveness of a workout. It also provides insight into a person's overall health, as changes in heart rate are a sign of potential health issues. Keeping track of a person's heart rate during physical activity is used to determine the amount of effort being put in, as well as how much rest is needed between sets. Additionally, changes in a person's heart rate indicate if they are over-exercising, which leads to injury or illness. Monitoring heart rate helps to detect any underlying health issues that may be present. In summary, tracking a person's heart rate while they exercise provides invaluable information, helping to ensure a safe and beneficial workout routine. However, some people may feel uncomfortable having their heart rate monitored while they exercise. Additionally, some people may feel that their privacy is being invaded if their heart rate data is being collected and stored.

- **Hand Hygiene Monitoring**

  This involves tracking how often and how effectively healthcare workers and patients wash their hands with soap and water. This helps to reduce the spread of infection, as well as providing a better understanding of the effectiveness of hand hygiene practices. This is important because if healthcare workers or patients do not wash their hands properly, they spread germs and bacteria from person to person. This leads to the spread of illnesses and infections, so it is important to monitor and track the hand hygiene practices of healthcare workers and patients. As such, it is essential to provide education and training about proper hand washing techniques and to ensure that hand hygiene practices are regularly and correctly followed. IoT devices are used by many hospitals today to remind patients to sanitize their hands when entering hospital rooms. The devices are even recommended how to sanitize in order to mitigate a particular risk for a specific patient. The main disadvantage of these devices is that they cannot cleanse hands for the user. Even so, research suggests these devices reduce hospital infection rates by over 60%.

- **Tracking of Depression and Moods**

  Depression is a state of low mood and aversion to activity. It affects a person's thoughts, behavior, feelings, and physical well-being. Tracking of depression and moods help individuals to identify potential triggers and inform treatment options to better manage symptoms and improve their overall well-being. This help individuals better understand their own mental health and recognize patterns in their behavior that may indicate a need for medical intervention. Additionally, tracking depression and moods helps clinicians identify any changes over time and adjust treatments accordingly. By using an IoT device, such as a fitness tracker or smartphone, to track individuals' moods, identifies patterns in their behavior that could point to depression or other mental health issues. With the data collected, individual work with a mental health professional to develop strategies to manage their moods or seek help if necessary. This provides a better understanding of how depression affects individuals over time and is used to develop better treatments or interventions for those suffering from the condition. These data are used to create personalized care plans, alert healthcare providers to sudden changes in mental state, and provide patients with real-time feedback and guidance on how to cope with their emotions.

- **Monitoring Parkinson's disease**

By monitoring Parkinson's disease progression, doctors better understand how the disease is affecting an individual and adjust treatment accordingly. This helps to ensure that the patient is receiving the best possible care and is managing their symptoms effectively. Monitoring disease progression also helps doctors understand how the disease is changing over time, which provides valuable insight into how to best treat the disease. This helps them identify new therapies and treatments that could be more effective at managing the disease than the treatments currently being used. An IoT device is used to monitor a person's movements and changes in their behavior that may indicate the onset of Parkinson's disease. It is used to monitor changes in medication and other lifestyle habits that have an impact on the disease. By tracking these changes, doctors and patients adjust therapies and treatments as needed to ensure the best possible outcome.

- **IoT-Based Connected Inhalers**

In conditions like asthma, attacks often occur suddenly and without warning. By using IoT-connected inhalers, healthcare providers can track the frequency of attacks and collect data on environmental factors that may have triggered them. This data is then used to create personalized treatment plans, helping patients identify and avoid triggers, and better manage their condition. The inhalers also alert healthcare providers when an attack is imminent, allowing for timely intervention if necessary. However, some argue that IoT-connected inhalers raise concerns about privacy invasion, potential overdiagnosis, and discrimination against patients with respiratory conditions.

- **IoT-Based Ingestible Sensors**

These sensors are small devices that are swallowed to track the body's internal processes. They measure a range of biometric data such as temperature, pH levels, and metabolic rate, and wirelessly transmit this data to a connected device. These sensors have the potential to revolutionize healthcare by enabling real time monitoring of a patient's health. They assist in diagnosing medical problems more quickly and accurately by detecting changes in body temperature, heart rate, and other vital signs. For instance, a temperature sensor embedded in a skin patch detect and alert doctors and caretakers about a patients' fever, prompting appropriate action. However, there are also potential dangers associated with these sensors. If the data collected by the sensors is not properly secured, it could be accessed by unauthorized individuals. This could lead to sensitive information about a patient's health being leaked. Additionally, if the sensors are not used properly, they could give false readings that could lead to misdiagnoses.

- **IoT-Based Connected Contact Lenses**

Connected contact lenses are contact lenses equipped with sensors and other technology that allow them to connect to other devices, such as smartphones. These lenses are embedded with tiny sensors that can detect physiological changes in the wearer's body and transmit that data to a connected device. This data is then used to monitor and manage various health conditions and provide more accurate diagnoses. By collecting this data and using it to inform diagnoses, healthcare providers are provided with more accurate and timely information, which leads to more effective treatments.

- **Robotic Surgery**

Robotic surgery, also known as robot-assisted surgery, involves the use of robotic assisted surgery, involves the use of robotics operated by a surgeon. This approach offers several advantages over traditional surgical techniques. The robotic arms are able to make incredibly precise movements, allowing the surgeon to make cuts and sutures with greater accuracy, which reduces the risk of tissue damage. The robotic arms also reduce the risk of infection, as they do not need to be touched by the surgeon and are operated from a distance. The surgeon is able to make precise movements that would not be possible with the human hand, which leads to better outcomes and reduced complications. However, it is important to note that the higher cost of robotic surgery poses a barrier for some patients in need [12].

9.5 Enabling Technologies in Healthcare for Sustainable Smart Cities

Technology-enabled healthcare is a transformative healthcare delivery in cities, offering enhanced access to healthcare services, involved health outcomes, and increased sustainability. This transformation is achieved through the utilization of innovative technologies, such as cloud computing, AI, and IoT, to improve the delivery of healthcare services and information. These technologies help improve the healthcare delivery system by allowing real-time data sharing and communication between healthcare providers, patients, and other stakeholders. This helps to improve efficiency, reduce costs, and create a more sustainable healthcare system. Additionally, it enables healthcare providers to use data-driven insights to better understand and predict patient needs and health trends. Some of the enabling technologies are discussed as follows:

9.5.1 Internet of Things

The Internet of Things is a technology that enables medical devices to collect, transmit, and analyze data in real time. This capability helps healthcare providers make better decisions and improve the quality of patient care. It is also utilized for remote patient monitoring, allowing patient to receive care without visiting the doctor's office. IoT facilitates the identification, detection, and authentication of medical and special care services. Globally, it refers to a number of physical devices that are connected to the internet to collect and share data. With the help of wireless sensors, the network gathers and shares data. Personalized, affordable, and cheap services are the hallmarks of IoT in digital healthcare. Using a wireless system, IoT helps doctors and nurses save time and effort. This allows medical practitioners to have real-time access to patient information, allowing them to make more informed decisions. Additionally, remote patient monitoring reduces the reliance on manual labor and resources, resulting in cost savings for the healthcare system.

According to the International Telecommunication Union (ITU), the IoT will move beyond anytime, anyplace connectivity for anyone to connectivity for anything, with initial emphasis on machine-to-machine (M2M) communication and digital identification [13]. Wireless Sensor Networks (WSN), Wireless Body Area Networks (WBAN), and Radio Frequency Identification (RFID) allow monitoring patient position and status [14–17]. It is possible to track chronic diseases with these wearable devices. These wearables devices, powered by AI, track steps, heart rate, blood pressure, calories, and more. The Wearable Internet of Things (WIoT) [18], Internet of Health Things (IoHT) [19], Internet of Medical Things (IoMT) [20], Internet of Nano Things (IoNT) [21] and Internet of mobile health Things (m-IoT) [22] are some of the variants derived from the core IoT concept in the healthcare fields depicted in Fig. 9.5.

9.5.2 Healthcare Cloud, Fog, and Edge Computing

Healthcare cloud computing is the use of cloud computing technologies in the healthcare industry. Cloud computing is a model for enabling ubiquitous, convenient, on-demand access to a shared pool of configurable resources,

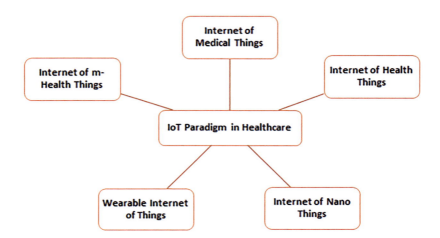

Fig. 9.5 IoT paradigm in healthcare

including computer networks, servers, storage, applications, and services. These resources are rapidly provisioned and released with minimal management effort or service provider interaction. It is a technology that allows for the storage, processing, and sharing of medical data in the cloud [23–27]. Cloud computing in healthcare is the use of remote computing resources for data storage and processing. It provides healthcare organizations with a secure, cost-effective way to store and manage large amounts of data. Additionally, it enables them to access data from anywhere and anytime and to collaborate on projects more easily. Moreover, it helps them to reduce information technology (IT) costs by eliminating the need for extra hardware and software, and allows them to scale up or down as needed. Furthermore, it enables them to collaborate more easily with other organizations, thereby leading to improved patient care. By leveraging cloud computing, healthcare organizations securely store and access vast amounts of data, achieving efficiency and cost-effectiveness in process manner.

Fog computing is a decentralized computing architecture that incorporates cloud computing into the IoT, enabling data processing closer to the source of data collection, rather than in a centralized cloud. It also facilitates faster and more secure data sharing with other organizations, leading to enhanced patient care. Fog computing provides the flexibility to scale data processing operations as needed, enabling quick adaptation to changing demands [28–31]. This approach allows healthcare organizations to improve response times, reduce latency, and alleviate the strain on the their data centers, resulting in cost savings. Edge computing, on other hand, reduces network latency by processing data locally instead of sending it to a cloud-based server. This reduces costs associated with data processing and enhance the security of patient data. Moreover, it enables the storage large amounts of data closer to the network edge, resulting in faster and more secure data processing [32, 33]. Healthcare cloud computing, fog computing, and edge computing collectively offer healthcare providers the ability to process large volumes of data rapidly, securely, and cost-effectively. These technologies play a crucial role in healthcare 4.0, positively impacting healthcare research and service improvement [34, 35] by enhancing quality, affordability and patient outcomes.

9.5.3 Big Data

Big data is a term used to describe large and complex data sets that are challenging to process using traditional data-processing applications. In the healthcare industry, big data refers to the collection, storage, and analysis of vast amounts of data to gain insights into patient care and health outcomes [36]. It has the potential to revolutionize the healthcare by enhancing patient care and improving efficiency. By leveraging data-driven analytics and ML, healthcare organizations gain insights into patient care that were previously unattainable. This assists in making informed decisions regarding treatment plans, diagnoses, and medication choices. Furthermore, big data in healthcare enables the identification of trends and patterns in patient care, leading to improved outcomes and reduced costs. Healthcare professionals are able to make more informed decisions, thereby enhancing the quality of care for patients. However, some experts are concerned that big data may lead to increased standardization in healthcare, potentially stifling creativity and innovation in the field. Moreover, critics argue that the use of big data in healthcare could have adverse effects. One concern is the potential for unfair targeting of specific groups, such as individuals with pre-existing conditions. Another worry is the possibility of manipulating patients into making decisions that may not align with their best interests. The widespread availability of information on the web, particularly through social networking sites, social media sources, and online discussion forums are dramatically changing medicine [37]. Real-time monitoring of individuals' health status, facilitated by stream processing systems, generates vast amounts of data from personal devices, wireless sensors, and mobile communication technologies. Additionally, medical tests, images, and clinical descriptions are recorded in records in various forms [38–40].

9.5.4 Blockchain Technology

Blockchain technology revolutionizes healthcare by providing a secure method for storing and transferring patient data. This technology creates a more secure, accurate, and transparent system for maintaining and sharing medical records and other health data. It reduces medical errors, enhances data security, and facilitates the tracking and analysis of patient health trends. Blockchain technology offers a secure and immutable solution for storing healthcare data, improving the accuracy of medical records, and reducing errors. It is used to store patient records, medical bills, clinical trial results, and other health data. Additionally, blockchain technology streamlines administrative tasks, reduces costs, and improves the quality of care. It also protects patient privacy by ensuring that only authorized users can access sensitive health information. One of the primary benefits of using blockchain technology in healthcare is the elimination of intermediaries, such as insurance companies, for mediating transactions between patients and healthcare providers. This reduces costs and enhances the security and transparency of transactions. Furthermore, it enables more efficient data sharing between healthcare providers, leading to more accurate diagnoses and effective treatments.

Data provenance, robustness, decentralized management, security, and privacy are considered key aspects of blockchain in health and medical services [41]. While blockchain technology has the potential to revolutionize healthcare, there are also risks and challenges associated with its adoption. For example, blockchain technology is still in its early stages of development, and its scalability is yet to be fully determined. Additionally, implementing blockchain technology in healthcare will require significant investment from organizations, and there is no guarantee of its success.

9.6 Smart Healthcare Applications

IoT-based smart healthcare applications enable remote monitoring of patients and provide real-time data to physicians and other healthcare providers. Here are a few examples of smart healthcare applications:

- **Drones**

  Drones provide a cost-effective way to deliver medical supplies to remote locations and are utilized for transporting samples and equipment for medical diagnoses. They also facilitate remote medical consultations, health conditions monitoring, and emergency response. Internet-based drones have various applications, including surveillance, mass testing, announcements, diagnostics, deliveries, and disinfection [42]. These drones are employed for delivering medical supplies to remote regions or aiding in potentially dangerous or hazardous areas. They ensure quick responses times during emergencies and offer medical care to those in need. Furthermore, they support surveillance and mass testing, enable medical announcements and diagnostics, and help disinfect areas to prevent the spread of diseases and infections. Diagnostic drones, such as thermal-imaging drones, infected can remotely monitor and detect infected cases by assessing their monitoring their temperature, heart rate, and symptoms like sneezing or coughing [43, 44].

- **Robotics Medical Applications**

  Robotics-based smart healthcare applications contribute to the improvement of diagnosis and treatment accuracy. They enhance patient monitoring and enable faster responses to medical emergencies. Moreover, they automate repetitive tasks, allowing healthcare workers to focus on more complex issues. For instance, humanoid robots minimize physical interactions between healthcare workers and infected individuals [45]. Robots assists with various tasks, including food preparation, medication administration, and cleaning and disinfection of medical facilities [46]. This streamlines processes, enhances efficiency and reduces the risk of human error. Additionally, robots are capable of performing hazardous tasks that may pose risk to humans, such as cleaning hazardous areas [47]. While robotics-based smart healthcare applications hold significant benefits, there are also potential risks. One concern is the possibility of excessive reliance on technology at the expense of human interaction. This could lead to a decline in the quality of patient care.

- **Assisted Living (In-Home Care)**

  Assisted living is a form of long-term care for elderly and disabled individuals who need help with daily activities, such as bathing, dressing, and medication management. It allows them to remain in their own home, while still receiving the care they need. This approach avoids unnecessary hospitalization by enabling patients to remain in their home environment [48]. Telepresence and videoconferencing robotic solutions provide a means to keep patients safe and monitored while also facilitating contact with family and healthcare professionals [49]. Moreover, these solutions are utilized to offer educational or therapeutic activities for older individuals, without necessitating the learning of complex technological systems [50]. Simpler and more ambient-assisted living solutions are designed to cater to individuals who may not be as familiar with technology, ensuring they can still benefit from its advantages [51]. The use of AI methodologies for ambient-intelligence systems in the healthcare domain are employed to process enormous amount of monitoring data efficiently scale with the number of ambient-assisted patients [52]. These technologies enable healthcare centers to receive automatic alerts for observation and emergency assistance [53]. Cloud and fog technologies are utilized to support healthcare and provide on-demand infrastructure [54–56].

- **Community and Children Healthcare Services**

The term "community and children's healthcare services" encompasses various healthcare services provided to children and families to improve their overall health and well-being. These services includes preventative care, primary care, mental health services, and more. These services offers a range of healthcare services for children in the local community, including preventative care, medical check-ups, vaccinations, and mental health support. Additionally, they provide education on nutrition and health topics, and offer access to resources for families in need. In the context of healthcare monitoring, an energy-efficient cooperative IoT platform has been proposed [57]. The concept of a community medical network is discussed in [58]. This approach aims to reduce the stigma associated with mental health issues and provide greater support for parents and children facing these challenges. It also ensure that children have access to the necessary resources for addressing any underlying mental health issues they may have [59]. An interactive totem is proposed for placement on a pediatric ward, providing services that educate, entertain, and empower hospitalized children [60]. IoT-based m-health services are presented to encourage children, along with their teachers and parents, to adopt healthy nutritional habits [61].

- **Personalized Healthcare**

Personalized healthcare involves the use of technology to individualize medical treatments for each patient [62–65]. This is done by collecting data on a patient's medical history, lifestyle, and genetics and then using that data to create a customized treatment plan that is tailored to the specific needs of each patient. By using this approach, healthcare professionals more accurately diagnose illnesses, identify potential risks, and design treatments that are tailored to the individual's unique needs. This leads to better patient outcomes and a better overall healthcare experience. By utilizing this personalized approach, healthcare professionals provide their patients with the most effective and individualized care possible, leading to

9.6 Smart Healthcare Applications

enhanced patient outcomes and satisfaction [66]. Thus, big data analytics is crucial for the implementation of individualized healthcare, both for individuals and population [67, 68]. In the context of personalized healthcare, big data analytics plays a crucial role in analyzing vast amounts of patient data to identify patterns, trends, and correlations. This analysis helps healthcare professionals make data-driven decisions, tailor treatment plans, and predict health outcomes for individual patients. By harnessing the power of big data analytics, personalized healthcare can be further enhanced, leading to improved patient care and outcomes. On the other hand, there are also some drawbacks to personalized healthcare. First, it can be expensive to implement, and second, it leads to healthcare providers becoming too reliant on technology, which could ultimately compromise patient care.

- **Rehabilitation**

Smart healthcare services enable patients to access personalized rehabilitation plans tailored to their individual needs. This enhances the effectiveness of their rehabilitation and reduces the required time for completion. Home-based rehabilitation is expected to reduce healthcare costs and improve patients' quality of life. IoT-based smart rehabilitation systems are deveoped using an ontology-based automated design method described in [69]. IoT-based technologies also support effective remote consultations in comprehensive rehabilitation [70]. As a branch of medicine that enhances and restores functional ability and quality of life for people with disabilities or impairments, physical medicine represents a vital field. Several IoT-based rehabilitation systems exist, including an integrated application system for prisons, a medical rehabilitation system for smart cities, and a language-training system for children with autism [62, 71–73].

- **Disease Monitoring, Telepathology, and Telemedicine**

Enabling technologies allow healthcare providers to monitor patients remotely, assess their medical condition, and provide advice without the patient having to visit a hospital or clinic. This reduces costs associated with healthcare and improves patient access to medical services. Remote patient monitoring systems and telemedicine have enabled healthcare providers to access patients' medical data remotely and provide health advice accordingly. This not only reduces the cost of healthcare but also increases the availability of medical services, even in remote areas. Telemedicine also helps improve the quality of medical care, as medical practitioners are able to access medical data from different locations and provide better diagnosis and treatment plans for their patients. This, in turn, helps to reduce medical costs as doctors are able to provide more efficient and accurate medical advice. Telepathology is a digital pathology technique that enables the remote viewing and interpretation of pathological images. It allows pathologists to collaborate and consult on cases regardless of their physical location, leading to improved access to expert opinions and faster diagnosis. Telepathology has the potential to enhance diagnostic accuracy, reduce turnaround time, and increase efficiency in healthcare systems.

- **Medication Management**

Medication management systems play a crucial role in ensuring accurate and safe delivery of medications to patients. They reduce errors, improve patient outcomes, and streamline the process of medication orders, administration, and dispensing. They also provide a way to document and store patient records, ensuring that the right medications are prescribed and administered. These systems also help to reduce the amount of time it takes for healthcare professionals to fill prescriptions, as well as streamlining the process of ordering and administering medications. These systems also automate processes, reducing costs, improving efficiency, and ensuring patients receive the correct medications in the correct dosages. Intelligent packaging of medicine boxes [74], and the use of RFID tags in an IoT-based medication management system further enhance medication control and monitoring [75].

- **Wheelchair Management**

Wheelchair management reduces the risk of falls and injuries, improving patient safety and experience. It also increases efficiency by minimizing patient transfer time and allowing medical staff to focus on other duties. Proper safety protocols, including engaging wheelchair brakes and securing patient, are essential. Trained staff should operate wheelchair and identify potential safety concerns. An IoT-based wheelchair healthcare system is proposed in [76], while [77] presents a peer-to-peer (P2P) and IoT-based medical support system. Monitoring vital signs and collecting environmental data, these technologies enable users to assess location accessibility.

- **Smartphones Services in Healthcare Systems**

Smartphone services are enabling healthcare systems to provide better patient care, monitor health status remotely, and increase patient engagement. This reduces costs, improve the quality of care, and increase access to healthcare. Through the use of mobile apps, patients can now easily access their health records, set up appointments, and communicate with their doctors. Moreover, these services enable remote monitoring of a patient's health status. These smartphone-controlled sensor services allow for patient monitoring, telemedicine, and remote doctor visits. They also enable medical professionals to access medical records quickly and securely, allowing for easier sharing of information between healthcare providers. Smartphones are used as versatile healthcare devices with the help of several hardware and software products. Information about diagnostics and treatments are accessed using diagnostic apps. Apps for medical education typically deal with tutorials, training, surgical demonstrations, color illustrations of various images, and medical books. Several image analysis algorithms that facilitate non-contact measurements for healthcare applications are presented in [78]. However, there are also several challenges that need to be addressed in order for IoT smartphones to be successfully implemented in healthcare systems. One of the main challenges is data security and privacy. With the increase in data breaches and cyberattacks, it is essential for healthcare organizations to have robust security measures in place to protect patient data. Another challenge is the lack of standardization among IoT devices. This lack of standardization makes it difficult for healthcare organizations to interoperate and exchange data between different devices and systems. However, not everyone is interested in fitness and health tracking, and some people may find it intrusive or even creepy to have their every move monitored by a device. In addition, for people who do want to track their fitness, there are many devices available that do not require a smartphone, such as Fitbit's Zip or One device.

9.7 Open Research Challenges and Research Directions

In recent years, researchers have been actively involved in designing and implementing healthcare services, as well as addressing various technical and architectural challenges. Alongside the research concerns in the literature, there are other open issues that require attention. These challenges encompass integrating diverse data sources, comprehending user behavior and usage patterns, developing personalized healthcare services, and advancing technologies for monitoring and diagnosis. Furthermore, research directions are aimed at developing predictive analytics, utilizing ML approaches for diagnosis, and employing visualization techniques for presenting healthcare data.

- **Policies and Standardization**

Policies and standardization help to improve the quality of healthcare services by providing a framework for the consistent delivery of care. They also help to reduce costs and errors and increase patient safety by ensuring that all healthcare providers are following the same protocols. It ensures that healthcare providers are delivering quality care in a consistent and cost-effective manner. Standardization helps to improve patient safety, reduce medical errors, streamline processes, and enhance the overall quality.

A company that develops e-Health policies may face challenges in adhering to standard rules and regulations for interoperability between devices. This is due to the complexity of managing different government policies, data privacy regulations, and the need to ensure standardized care delivery across various providers. When defining standards, it is important to consider a wide range of topics, including communications layers and protocol stacks.

Various value-added services, such as electronic health records, also require standardization. To achieve this, organizations such as the Information Technology and Innovation Foundation (IETF), the Internet Protocol for Smart Objects (IPSO) alliance, and the European Telecommunication Standards Institute (ETSI) collaborate through IoT technology working groups. These groups work alongside m-Health and e-Health organizations and IoT researchers to standardize IoT-based healthcare services. Additionally, the healthcare sector is fragmented, with different stakeholders having diverse interests and objectives, makes implementing a unified policy challenging. Despite these challenges, policies and standardization in healthcare offer numerous benefits. They help ensure that patients receive consistent, high quality care regardless of the treatment location. Furthermore, these policies enhance communication and coordination among healthcare providers, leading to improved patient outcomes.

- **Security and Privacy**

Security and privacy are crucial in healthcare services to ensure the safety and confidentiality of patient data, as well as to prevent the unauthorized sharing of sensitive information. It must ensure that patient data remains secure and private, as mandated by HIPAA regulations. This involves protecting patient data from unauthorized access and ensuring that only authorized personnel can access patient information. HIPAA regulations require healthcare service providers to implement a variety of measures to protect patient data, such as encryption and access control systems. Healthcare providers must also have a written policy in place to ensure that patient data is protected and that any breaches of the policy are reported immediately. In addition, data security measures must be in place to protect against unauthorized access to patient records. In order to ensure the confidentiality of medical information, unauthorized users are unable to access it. An IoT health device uses authentication to verify the identity of its peers. A network service or resource is accessible only to authorized nodes when they are authorized. Even if a fault occurs, a security scheme should continue to provide its respective security services. Healthcare services collect, store, and share a lot of sensitive and personal data, making them vulnerable to cyberattacks. In addition, due to the complexity of modern healthcare systems, it is difficult to ensure that the data is secure and only accessible to authorized personnel.

These challenges have a major impact on the quality of patient care and must be addressed with effective policies and procedures. For instance, patient data must be stored in secure databases that are regularly updated for the latest security patches and monitored for any suspicious activities. However, some have argued that these challenges are not as significant as they seem and that the benefits of digital health outweigh the risks. For example, digital health can help to improve patient outcomes by providing real-time data that is used to make decisions about treatment, as well as increasing transparency and communication between patients and providers.

- **Monitoring, Mobility, and Connectivity**

Monitoring, mobility, and connectivity are essential elements of healthcare, as they enable services to be accessible to more people and facilitate tracking and monitoring of patient health. This allows healthcare providers to monitor their patients from anywhere and provide better care. Connectivity also enables patients to remotely access their medical records and data, making it more convenient to seek medical advice and treatments. Additionally, mobility facilitates healthcare providers to reaching more people, as they are not limited to being in the same location as their patients. In a connected smart health system, both patients and medical staff should have the ablility to move freely between multiple locations. However, the presence of several wireless networks operating in close proximity leads to collisions, particularly when there is mobility involved. Healthcare delivery faces disruptions, and reduced performance, which may lead to disastrous situations. Therefore, its is crucial to ensure the accurate operation of medical devices by connecting them through various wireless technologies. The challenges associated with monitoring, connectivity, and mobility in IoT/mobile applications are numerous. Despite significant progress in sensors and IoT technology, achieving the same level of accuracy as hospital-grade devices while maintaining energy efficiency and wearability remains a challenge. To address this, we need new technologies that are cost-aware, powerful, fully connected, and efficient. The evolution and innovation of smart and connected healthcare are expected to provide efficient solutions and services to patients, paving the way for next-generation healthcare. Research and development efforts are necessary to ensure data security and user privacy. Additionally, there is a need for research to improve methods for monitoring healthcare services and enhancing their mobility and connectivity.

9.8 Summary

Across the globe, researchers, scientists, industries, and government are exploring the healthcare sector, which has attracted their attention. Smart healthcare involves using technology to improve the quality of healthcare delivery, from diagnosis to treatment. It includes artificial intelligence, virtual reality, and other advanced technologies to streamline the healthcare system and increase efficiency. In the Fourth Industrial Revolution, key focuses in healthcare are Internet of Things, cloud and fog computing, and big data analytics. These advancements help reduce healthcare costs by minimizing the number of necessary visits for diagnosis and treatment. They also contribute to improving the accuracy of diagnosis and treatment, leading to fewer complications and better patient outcomes. By reducing the time and money spent on diagnosing and treating illnesses, smart and connected healthcare technologies yields improved health outcomes for patients. By providing instant access to medical records, doctors quickly and accurately diagnose illnesses and prescribe personalized treatment plans. This result in fewer complications and hospital visits, reducing costs for patients and the healthcare system as a whole. Immediate access to patient data helps healthcare professionals identify the most effective treatment options, saving time and money by avoiding ineffective treatments and achieving better outcomes.

References

1. S.P. Mohanty, U. Choppali, E. Kougianos, Everything you wanted to know about smart cities: The Internet of Things is the backbone. IEEE Consum. Electron. Mag. **5**(3), 60–70 (July 2016)
2. J.L. Martin, H. Varilly, J. Cohn, G.R. Wightwick, Preface: Technologies for a smarter planet. IBM J. Res. Dev. **54**(4), 1–2 (2010)
3. F.F. Gong, X.Z. Sun, J. Lin, X.D. Gu, Primary exploration in establishment of China's intelligent medical treatment. Mod Hosp. Manag. **11**(2), 28–29 (2013)
4. K. Ullah, M.A. Shah, S. Zhang, Effective ways to use Internet of Things in the field of medical and smart health care, in *Proceedings under International Conference on Intelligent Systems Engineering (ICISE)*, (2016), pp. 372–379
5. M. Carolyn Clancy, Getting to 'smart' health care: Comparative effectiveness research is a key component of, but tightly linked with, health care delivery in the information age. Health Aff. **25**(1), 589–592 (2006)
6. Smart and Connected Health (SCH) Program Solicitation NSF 18-541, National science foundation, Alexandria, VA, USA (2013). Available on: https://www.nsf.gov/pubs/2018/nsf18541/nsf18541.htm. Accessed on 24 Mar 2022
7. M. Chen, J. Qu, Y. Xu, J. Chen, Smart and connected health: What can we learn from funded projects? Data Inf. Manag. **1**(2), 141–152 (2018)
8. A Guide to the Internet of Things Infographic, Intel. Available on: https://www.intel.com/content/www/us/en/internet-of-things/infographics/guide-to-iot.html. Accessed 13 Nov 2022
9. 87% of healthcare organizations will adopt internet of things technology by 2019, HIPAA J (2017). Available on: https://www.hipaajournal.com/87pc-healthcare-organizations-adopt-internet-of-things-technology-2019-8712/. Accessed 6 Mar 2023
10. R.S.H. Istepanian, The potential of Internet of Things (IoT) for assisted living applications, in *Proceedings of IET Seminar Assisted Living*, (2011), pp. 1–40
11. R.S.H. Istepanian, S. Hu, N.Y. Philip, A. Sungoor, The potential of Internet of m-health Things 'm-IoT' for non-invasive glucose level sensing, in *Proceedings of IEEE Annual International Conference of the IEEE Engineering in Medicine and Biology Society (EMBC)*, (2011), pp. 5264–5266
12. 10 Internet of Things (IoT) Healthcare examples, and why their security matters. Available on: https://ordr.net/article/iot-healthcare-examples/. Accessed on 8 Mar 2023
13. L. Atzori, A. Iera, G. Morabito, The internet of things: A survey. Comput. Netw. **54**(15), 2787–2805 (2010)
14. E. Ilie-Zudor, Z. Kemny, F. van Blommestein, L. Monostori, A. van der Meulen, A survey of applications and requirements of unique identification systems and rfid techniques. Comput. Ind. **62**(3), 227–252 (2011)
15. P. Rawat, K.D. Singh, H. Chaouchi, J.M. Bonnin, Wireless sensor networks: A survey on recent developments and potential synergies. J. Supercomput. **68**(1), 1–48 (2014)
16. H. Cao, V. Leung, C. Chow, H. Chan, Enabling technologies for wireless body area networks: A survey and outlook. IEEE Commun. Mag. **47**(12), 84–93 (Dec 2009)
17. M. Chen, S. Gonzalez, A. Vasilakos, H. Cao, V.C. Leung, Body area networks: A survey. Mob. Netw. Appl. **16**(2), 171–193 (2011)
18. S. Hiremath, G. Yang, K. Mankodiya, Wearable internet of things: Concept, architectural components and promises for person-centered healthcare, in *Proceedings of Wireless Mobile Communication and Healthcare (Mobihealth), 2014 EAI 4th International Conference on*, (IEEE, 2014), pp. 304–307
19. N. Terry, Will the internet of things disrupt healthcare? Vand. J. Ent. Tech. L. **19**, 327 (2016)
20. N.K. Jha, Internet-of-medical-things, in *Proceedings of the on Great Lakes Symposium on VLSI 2017, GLSVLSI'17*, (New York, NY, USA, 2017)
21. E. Omanovic-Miklicanin, M. Maksimovic, V. Vujovic, The future of healthcare: Nanomedicine and internet of nano things. Folia Med. Facult. Med. Univ. Saraeviensis **50**(1), 23–28 (2015)

References

22. T.N. Gia, IoT-based continuous glucose monitoring system: A feasibility study. Proc. Comput. Sci. **109**, 327–334 (2017)
23. P.D. Kaur, I. Chana, Cloud based intelligent system for delivering health care as a service. Comput. Methods Prog. Biomed. **113**(1), 346–359 (2014)
24. H. Shafiei, A. Khonsari, P. Mousavi, Serverless computing: A survey of opportunities, challenges, and applications. ACM Comput. Surv. **54**(11s), 1–32 (2022)
25. A. Botta, W. de Donato, V. Persico, A. Pescape, Integration of cloud computing and internet of things: A survey. Futur. Gener. Comput. Syst. **56**, 684–700 (2016)
26. L. Atzori, A. Iera, G. Morabito, Understanding the internet of things: Definition, potentials, and societal role of a fast evolving paradigm. Ad Hoc Netw. **56**, 122–140 (2017)
27. J. Schlechtendahl, M. Keinert, F. Kretschmer, A. Lechler, A. Verl, Making existing production systems industry 4.0-ready. Prod. Eng. **9**(1), 143–148 (2015)
28. T.N. Gia, M. Jiang, A.-M. Rahmani, T. Westerlund, P. Liljeberg, H. Tenhunen, Fog computing in healthcare internet of things: A case study on ecg feature extraction, in *Computer and Information Technology, Ubiquitous Computing and Communications, Dependable, Autonomic and Secure Computing, Pervasive Intelligence and Computing (CIT/IUCC/DASC/PICOM), 2015 IEEE International Conference on*, (IEEE, 2015), pp. 356–363
29. L.M. Vaquero, L. Rodero-Merino, Finding your way in the fog: Towards a comprehensive definition of fog computing. SIGCOMM Comput. Commun. Rev. **44**(5), 27–32 (2014)
30. A. Kumari, S. Tanwar, S. Tyagi, N. Kumar, Fog computing for healthcare 4.0 environment: Opportunities and challenges. Comput. Electr. Eng. **72**, 1–3 (2018)
31. A.V. Dastjerdi, R. Buyya, Fog computing: Helping the internet of things realize its potential. Computer **49**(8), 112–116 (2016)
32. M. Etsi, *Mobile-edge computing* (Introductory Technical White Paper, 2014)
33. N. Fernando, S.W. Loke, W. Rahayu, Mobile cloud computing: A survey. Futur. Gener. Comput. Syst. **29**(1), 84–106 (2013)
34. F. Andriopoulou, T. Dagiuklas, T. Orphanoudakis, *Integrating iot and fog computing for healthcare service delivery* (Springer, Components and Services for IoT Platforms, 2017)
35. Y. Shi, G. Ding, H. Wang, H.E. Roman, S. Lu, The fog computing service for healthcare, in *Future Information and Communication Technologies for Ubiquitous HealthCare (Ubi-HealthTech), 2015 2nd International Symposium on IEEE*, (2015), pp. 1–5
36. J. Gantz, D. Reinsel. *Extracting Value from Chaos*. Technical Report 2011 (2011)
37. N. Waxer, D. Ninan, A. Ma, N. Dominguez, How cloud computing and social media are changing the face of health care. Physician Exec. **39**(2), 58 (2013)
38. S. Ullah, H. Higgins, B. Braem, B. Latre, C. Blondia, I. Moerman, S. Saleem, Z. Rahman, K.S. Kwak, A comprehensive survey of wireless body area networks. J. Med. Syst. **36**(3), 1065–1094 (2012)
39. D. Garets, M. Davis, *Electronic medical records vs. electronic health records: Yes, there is a difference*, Policy white paper (HIMSS Analytics, Chicago, 2006), pp. 1–4
40. K. Smolij, K. Dun, Patient health information management: Searching for the right model. Perspect. Health Inf. Manag. **3**(10), 1–11 (2006)
41. T.T. Kuo, H.E. Kim, L. Ohno-Machado, Blockchain distributed ledger technologies for biomedical and health care applications. J. Am. Med. Inform. Assoc. **24**, 1211–1220 (2017)
42. L. Sedov, A. Krasnochub, V. Polishchuk, Modeling quarantine during epidemics and mass-testing using drones. PLoS One **24**(6), e0235307/1–11 (2020)
43. T. Cozzens, Pandemic Drones to Monitor, Detect Those With COVID-19_GPS World?: GPS World (2020). Accessed 16 Mar 2021. [Online]. Available: https://www.gpsworld.com/dragan_y-camera-anduav-expertise-to-help-diagnose-corona virus/ Accessed on 6 Mar 2023
44. A. Kumar, K. Sharma, H. Singh, S.G. Naugriya, S.S. Gill, R. Buyya, A drone-based networked system and methods for combating coronavirus disease (COVID-19) pandemic. Futur. Gener. Comput. Syst. **115**, 1–29 (2020)
45. Z. Zeng, P.J. Chen, A.A. Lew, From high-touch to high-tech: COVID-19 drives robotics adoption. Tour. Geogr. **22**(3), 724–734 (2020)

46. Z.H. Khan, A. Siddique, C.W. Lee, Robotics utilization for healthcare digitization in global COVID-19 management. Int. J. Environ. Res. Public Health **17**(11), 3819/1–21 (May 2020)
47. M. Podpora, A. Gardecki, R. Beniak, B. Klin, J.L. Vicario, A. Kawala-Sterniuk, Human interaction smart Subsystem Extending speech-based human-robot interaction systems with an implementation of external smart sensors. Sensors **20**(8), 2376/1–16 (2020)
48. V. Osmani, S. Balasubramaniam, D. Botvich, Human activity recognition in pervasive healthcare: Supporting efficient remote collaboration. J. Netw. Comput. Appl. **31**(4), 628–655 (2008)
49. S. Patel, H. Park, P. Bonato, L. Chan, M. Rodgers, A review of wearable sensors and systems with application in rehabilitation. J. Neuroeng. Rehabil. **9**(1), 1–7 (2012)
50. T.S. Dahl, M.N.K. Boulos, Robots in health and social care: A complementary technology to home care and telehealthcare? Robotics **3**(1), 1–21 (2013)
51. L.G. Yang, M. Xie, X. Mantysalo, Z. Zhou, L. Pang, S. Da Xu, Q.C. Kao-Walter, L.-R. Zheng, A health-iot platform based on the integration of intelligent packaging, unobtrusive bio-sensor, and intelligent medicine box. IEEE Trans. Industr. Inform. **10**(4), 2180–2191 (2014)
52. G. Acampora, D.J. Cook, P. Rashidi, A.V. Vasilakos, A survey on ambient intelligence in healthcare. Proc. IEEE **101**(12), 2094–2470 (2013)
53. H. Xia, I. Asif, X. Zhao, Cloud-ecg for real time ecg monitoring and analysis. Comput. Methods Prog. Biomed. **110**(3), 253–259 (2013)
54. M. Chen, Y. Ma, S. Ullah, W. Cai, E. Song, Rochas: Robotics and cloud assisted healthcare system for empty nester, in *Proceedings of the 8th International Conference on Body Area Networks, ICST (Institute for Computer Sciences, Social- Informatics and Telecommunications Engineering)*, (2013), pp. 217–220
55. M. Deng, M. Petkovic, M. Nalin, I. Baroni, A home healthcare system in the cloud–addressing security and privacy challenges, in *Proceedings of Cloud Computing (CLOUD), 2011 IEEE International Conference on IEEE*, (2011), pp. 549–556
56. D. Gachet, M. de Buenaga, F. Aparicio, V. Padron, Integrating internet of things and cloud computing for health services provisioning: The virtual cloud carer project, in *Proceedings of Innovative Mobile and Internet Services in Ubiquitous Computing (IMIS), 2012 Sixth International Conference on IEEE*, (2012), pp. 1–6
57. V.M. Rohokale, N.R. Prasad, R. Prasad, A cooperative Internet of Things (IoT) for rural healthcare monitoring and control, in *Proceedings of International Conference on Wireless Communication, Vehicular Technology, Information Theory and Aerospace & Electronic Systems Technology (Wireless VITAE)*, (2011), pp. 1–6
58. L. You, C. Liu, S. Tong, Community medical network (CMN): Architecture and implementation, in *Proceedings of Global Mobile Congress (GMC)*, (2011), pp. 1–6
59. Awareness Day 2014 Activities by Program Type. [Online]. Available: http://www.samhsa.gov/sites/default/_les/children-awareness-dayactivities-by-program-2014.pdf. Accessed 7 Dec 2022
60. S. Vicini, S. Bellini, A. Rosi, S. Sanna, An Internet of Things enabled interactive totem for children in a living lab setting, in *Proceedings of ICE International Conference on Intelligent Innovations (ICE)*, (2012), pp. 1–10
61. M. Vazquez-Briseno, C. Navarro-Cota, J.I. Nieto-Hipolito, E. Jimenez-Garcia, J.D. Sanchez-Lopez, A proposal for using the Internet of Things concept to increase children's health awareness, in *Proceedings of 22nd International Conference on Communication, Computing (CONIELECOMP)*, (2012), pp. 168–172
62. S.M.R. Islam, D. Kwak, M.H. Kabir, M. Hossain, K.S. Kwak, The internet of things for health care: A comprehensive survey. IEEE Access **3**, 678–708 (2015)
63. N. Surantha, P. Atmaja, M. Wicaksono, A review of wearable internet-of-things device for healthcare. Procedia Comput. Sci. **179**, 936–943 (2021)
64. M. Viceconti, P. Hunter, R. Hose, Big data, big knowledge: Big data for personalized healthcare. IEEE J. Biomed. Health Inform. **19**(4), 1209–1215 (2015)
65. M. Maksimovic, The roles of Nanotechnology and Internet of Nano things in healthcare transformation. TecnoLogicas **20**(40), 139–153 (2017)

References

66. N.T. Issa, S.W. Byers, S. Dakshanamurthy, Big data: The next frontier for innovation in therapeutics and healthcare. Expert. Rev. Clin. Pharmacol. **7**(3), 293–298 (2014)
67. R. Chen, M. Snyder, Promise of personalized omics to precision medicine. Wiley Interdiscip. Rev. Syst. Biol. Med. **5**(1), 73–82 (2013)
68. N.V. Chawla, D.A. Davis, Bringing big data to personalized healthcare: A patient-centered framework. J. Gen. Intern. Med. **28**(3), 660–665 (2013)
69. Y.J. Fan, Y.H. Yin, L.D. Xu, Y. Zeng, F. Wu, IoT-based smart rehabilitation system. IEEE Trans. Ind. Inf. **10**(2), 1568–1577 (May 2014)
70. B. Tan, O. Tian, Short paper: Using BSN for tele-health application in upper limb rehabilitation, in *Proceedings of IEEE World Forum Internet Things (WF-IoT)*, (2014), pp. 169–170
71. D.Y. Lin, Integrated Internet of Things application system for prison, *Chinese Patent 102 867 236 A*, (January 9, 2013)
72. L. Fan, Usage of narrowband internet of things in smart medicine and construction of robotic rehabilitation system. IEEE Access **10**, 6246–6259 (2021)
73. G.S. Karthick, P.B. Pankajavalli, A review on human healthcare internet of things: A technical perspective. SN Comput. Sci. **1**(4), 198/1–19 (2020)
74. Z. Pang, J. Tian, Q. Chen, Intelligent packaging and intelligent medicine box for medication management towards the Internet-of-Things, in *Proceedings of 16th International Conference on Advanced Communication Technology (ICACT)*, (2014), pp. 352–360
75. I. Laranjo, J. Macedo, A. Santos, Internet of Things for medication control: E-health architecture and service implementation. Int. J. Rel. Q. E-Healthc. **2**(3), 1–15 (2013)
76. L. Yang, Y. Ge, W. Li, W. Rao, W. Shen, A home mobile healthcare system for wheelchair users, in *Proceedings of IEEE International Conference on Computer Supported Cooperative Work Design (CSCWD)*, (2014, May), pp. 609–614
77. V. Kolici, E. Spaho, K. Matsuo, S. Caballe, L. Barolli, F. Xhafa, Implementation of a medical support system considering P2P and IoT technologies, in *Proceedings of 8th International Conference on Complex, Intelligent and Software Intensive Systems (CISIS)*, (2014), pp. 101–106
78. P.J.F. White, B.W. Podaima, M.R. Friesen, Algorithms for smartphone and tablet image analysis for healthcare applications. IEEE Access **2**, 831–840 (2014)

Chapter 10
Unmanned Aerial Vehicles in Sustainable Smart Cities

10.1 Introduction

The world's population is rapidly increasing and expected to double by 2050 with a constant migration from rural to urban areas. In the coming years, this trend is expected to grow even more rapidly, particularly in developing countries. City administrators aiming to maintain or enhance the quality of life for city residents, will face many challenges due to the rapid population growth in cities. Therefore, smart cities developments are increasingly incorporating advanced information communication technology (ICT), robotics, and smart solutions [1]. The use of these technologies will improve infrastructure performance and increase residents' comfort levels by creating smart automated services. In smart cities, infrastructures and services are designed to be efficient while reducing costs. There has been a significant improvement in the design and abilities of unmanned aerial vehicles (UAVs) due to advances in robotics. It is now possible to purchase inexpensive UAVs equipped with microprocessors, data storage, sensors, actuators, cameras, and wireless communication devices. Despite their reliability, they offer a very high level of flight stability. A number of civil applications have recently emerged, contributing to further improving UAV capabilities at a lower cost compared to a few years ago. A typical UAV consists of several components, including a flight system, a control system, a monitoring system, a data processing system, and a landing system. In most cases, an on-board system handles navigation, data gathering, and data transmission.

After UAVs were allowed to be employed in civil applications, their image began to change. For example, UAVs were involved in humanitarian operations where they assisted in finding people who needed help and played a crucial role in saving lives during catastrophic situations. They were also utilized as first responder tools to assess the situation, locate survivors, determine the extent of assistance required, and provide support. Furthermore, UAVs were employed for wildlife protection. In Nepal, guards were trained to operate UAVs to protect wildlife, leading to significant improvements in addressing major issues. Japan utilizes UAVs to monitor illegal Japanese whaling in the southern hemisphere. These applications, among others, have created positive impressions of UAVs and encourage their widespread use [2]. In addition, UAVs can contribute significantly and effectively to the development of other civilian applications such as surveying activities, environmental monitoring, and traffic control.

In smart cities, UAVs are capable of being used for a wide range of applications due to their flexibility and speed. These applications include traffic and crowd monitoring, environmental monitoring, civil security, and merchandise delivery. Smart cities will benefit from UAV applications designed for them as illustrated in Fig. 10.1. Using UAVs equipped with sensors and cameras to monitor a city's infrastructure and employing software to analyze the collected data, users are capable of effectively inspecting and controlling smart cities' infrastructure. Initially, UAVs were primarily used for military purposes, however, in recent years a diverse range of civil applications has been developed, including agriculture, logistics, environmental protection, public safety, and traffic control. Another emerging application

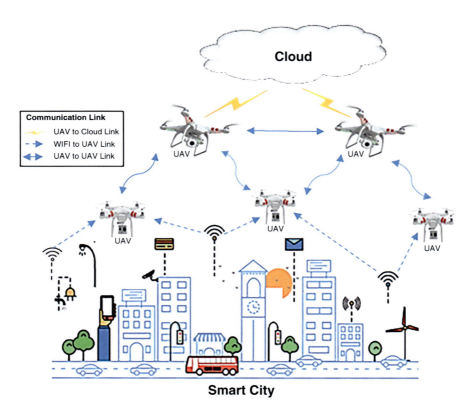

Fig. 10.1 Applications of UAV in the sustainable smart cities

of UAVs is their involvement in smart cities, which presents several benefits. The cost effectiveness of UAVs offers significantly higher level than that of manned aircraft. As a result, they are more adaptable to various environments and situations, including those that are challenging and pose high risks to humans. Moreover, UAVs fly in close proximity to target objects, improving measurement accuracy and targeting.

Most UAV applications require mobility and access to places that are difficult to reach [3]. These applications are typically carried out by manned airplanes or helicopters, which involves repetitive work or hazardous conditions. However, relying on manned aircraft results in significant costs and overhead. UAVs offers a more efficient and cost-effective alternative for performing such tasks. They are known for their high accuracy, flexibility, repeatability, and mobility [4]. Moreover, the deployment of manned planes for risky tasks may end up putting pilots at risk, whereas UAVs minimizes or eliminate this risk. There is a rapid increase in the demand for UAVs, and continuous development is underway for various advanced applications. It is anticipated that many public and private organizations will extensively utilize UAVs to enhance the quality, productivity, flexibility, and cost of their services [5]. The purpose of this chapter is to provide insights into the potential applications and challenges associated with integrating UAVs into smart cities. It also analyzes the ethical considerations surrounding UAV use, such as privacy concerns and the responsible deployment of unmanned aerial systems. By addressing these multidimensional aspects, this chapter provides a comprehensive overview of the opportunities and obstacles associated with the integration of UAVs, facilitating informed decision-making and fostering the development of truly transformative smart cities. By understanding these applications and challenges, stakeholders make informed decisions and harness the full potential of UAV technology in building smarter and more sustainable cities.

10.2 Role of UAVs in Sustainable Smart Cities

Smart cities have existed for decades, but their latest version places a greator emphasis on technology, particularly advanced robotics, in order to improve living conditions for citizens. The aim is to enhance sustainability by integrating cutting-edge technology with efficient infrastructure. With the rise of urban populations, the expenditure required to meet the demands of these cities will rise as well. It is estimated that smart city technology will cost nearly $135 billion by 2021, according to a study from the International Data Corporation. The use of unmanned aerial vehicles (UAVs, or drones) is likely to have a significant and diverse impact on the evolution of smart cities, despite the fact that a number of emerging technologies will be considered. In a recent report published in Science Direct, UAVs were found to provide cost-effective services to local governments, from environmental monitoring to traffic management.

With modest technological and security advances, UAVs have found extensive applications in cities, including traffic and crowd monitoring, civil security, merchandise delivery, infrastructure inspection, and more. The comprehensive use of UAVs in day-to-day municipal operations offers advantages that align with the objectives of smart cities, sets out to enhance the lives of residents. Traffic congestion has long been a challenge for cities worldwide, from sudden gridlocks to predictable rush hour traffic. In order to find a reasonable solution to traffic woes, it is crucial that cities first understand basic information about the causes of congestion. This information includes the status of the roads in the most crowded areas, peak hours, public transport availability, etc. While static cameras provide limited intelligence in collecting this information, UAVs play an immediate and crucial role in managing traffic within cities. Equipped with the necessary tools, UAVs collect and deliver real-time traffic congestion data. With UAV's high visibility and maneuverability, it provides live feeds of congestion, enabling other technologies or human intervention to address the traffic issue promptly or redirect traffic to avoid further congestion. When combined with other ICT or Internet of Things (IoT) innovations, UAVs assist drivers in finding parking spaces, reducing vehicle miles traveled for similar activities. Moreover, drones contribute to mapping out metro projects, optimizing bus routes, and identifying suitable places for bike paths and other sustainable transportation systems. Some of the potential applications of UAVs in various components of smart cities is given in Table 10.1.

UAVs plays a crucial role in monitoring existing congestion and preventing future congestion through their combined efforts. Their consistent and comprehensive monitoring capabilities also assist authorities in taking precautionary measures in the face of natural disasters. In a natural situations where humans safety is at risk, UAVs serves as reliable, flexible, and safe tools for deploying messages and analyzing real-time circumstances. The same applies to the deployment of emergency health services, where UAVs greatly enhance cities' response to public health needs. They enable the quick delivery of medical supplies and services and serves as functional ambulatory services, providing essential life support equipment and medical supplies. The benefits of UAVs for smart cities are extensive, encompassing areas such as smart transportation, crowd management, tourism support, and merchandise delivery. However, the realization of these smart solutions and wide-ranging applications is contingent upon overcoming several challenges. Currently, the integration of UAVs into smart cities is hindered by issues related to licensing, certification, privacy and security concerns. To unlock the potential of a "smart city" by leveraging UAVs applications, government leaders must address the associated risks and concerns.

In preliminary data analysis, Fig. 10.2a illustrates the number of documents published per year with the number of publications. The most publications are made in the year 2021 with 35, and as publications are growing each year, this is an excellent field for further research. Articles are the document type with a large number of publications with a percentage of 60.8%. Figure 10.2b shows the document per year by source with subject area. IEEE Access publishes a wide range of scholarly articles, with an extensive collection of over 6 publications in 2019. Computer Science

10.3 Technologies Used for UAVs in Sustainable Smart Cities

Table 10.1 Applications of UAV in various components of smart cities

Smart city components	Description	References
Smart economy	Photogrammetry	[6, 7]
	Urban and peri-urban agriculture	[8–10]
	Engineering and construction	[11, 12]
Smart governance	Urban planning	[13]
	Species monitoring	[14]
	Pandemic management	[15, 16]
	Big data	[17]
	Infrastructure inspection (buildings, bridges, and grid)	[18, 19]
Smart environment	Environment and water monitoring	[20–22]
Smart living	Crowd management	[23]
	Disaster management	[24–26]
	Medical UAS	[27, 28]
	Green housing	[29]
	Journalism	[30, 31]
	Tourism	[32, 33]
	UAS for filming sport and cultural events	[34, 35]
Smart people	Education	[36, 37]
	Behavioral mapping	[38]
Smart mobility	Traffic monitoring and management	[39–41]
	Urban air mobility	[42, 43]
	Smart parking	[44]

subject domain has most of the publications with a percentage of 34.8%, and Engineering has the second-most publications with a percentage of 25.9%. Figure 10.2c shows the congregation of keywords that are occurring together in research papers. The interrelations between keywords that are co-appearing in the research are linked together. Unmanned aerial vehicles, Smart City, Internet of Things, drones, and antennas are the most co-appearing keywords.

The rest of the chapter is organized as follows. The role of UAVs in sustainable smart cities is presented in Sect. 10.2. The technologies and applications of UAVs in sustainable smart cities are discussed in Sects. 10.3 and 10.4, respectively. The open research challenges and research directions are discussed in Sect. 10.5. Finally, Sect. 10.6 concludes with the summary.

10.3 Technologies Used for UAVs in Sustainable Smart Cities

Technologies are applied to a smart city's components and services to make it smart. In this section, we discuss various technologies used for UAV in sustainable smart cities as shown in Fig. 10.3.

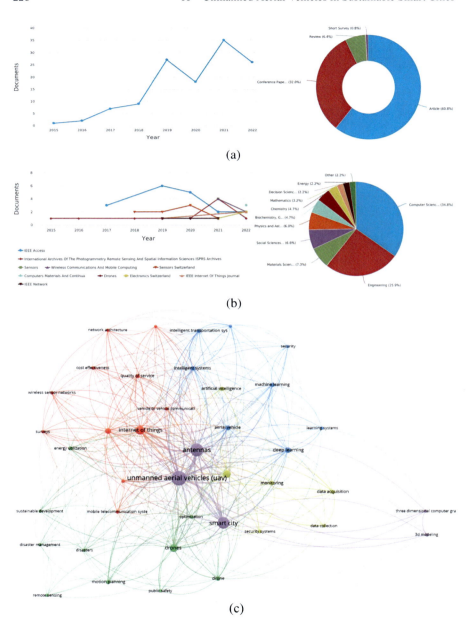

Fig. 10.2 Preliminary data analysis (**a**) documents published per year with the number of publications, (**b**) document per year by source with subject area, (**c**) the congregation of co-appearing keywords from the Scopus database

10.3 Technologies Used for UAVs in Sustainable Smart Cities

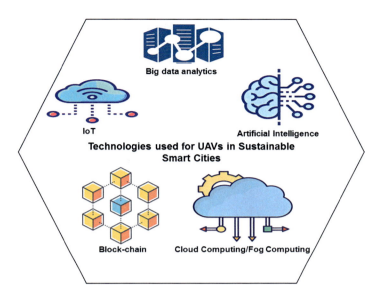

Fig. 10.3 Technologies used for UAVs in sustainable smart cities

10.3.1 Internet of Things (IoT)/Internet of Drone (IoD)

A UAV-based IoT system will play an essential role in future wireless networks due to the flexibility of UAV deployment and the dominance of Line of Sight (LoS) channels. In a UAV-based IoT system [45], flexible UAVs are responsible for collecting data from IoT devices. These UAVs provide reliable data transmission services, leveraging the dominant LoS, which also overcomes geometrical limitations. An Internet of Everything (IoE) ecosystem is designed to utilize UAVs for data dissemination. When spectrum becomes available, these UAVs make a decision using the provided algorithm. A joint optimization of the resource assignment strategy is designed for efficient and flexible data dissemination between IoT devices [46]. The closed-loop transmission diversity approach was developed to enhance the transmission of data over low-power devices across long distances. The UAVs must be scheduled, bandwidth allocated, and mobile to maximize data reception from the base stations.

A subset of the IoT that is certainly considered is the Internet of Drones (IoD). Essentially, it provides controlled and coordinated access to supervised airspace for UAVs, also known as drones [47]. These drones are rapidly finding novel and productive uses as sensors, processors, cameras, digital memory, and ubiquitous wireless connectivity continue to decrease in size. This technology is used for various purposes such as military surveillance, disaster surveillance, package delivery, traffic and wildlife monitoring, agriculture, infrastructure inspection, bridges, tunnels, cinematography, and more. Having a proper navigation system and airspace management for drones is imperative. To make IoD a reality, drone simulations are necessary to ensure their safety and reliability. It is crucial that they are low-cost and easy to maintain, allowing for widespread acceptance and use. Reliable software or apps must be certified for safety, adhere to effective communication protocols, and be energy efficient. In challenging environments, they should be designed to function reliably and effectively despite adverse conditions or constraints [48].

10.3.2 Artificial Intelligence (AI)/Machine Learning (ML)

Human intelligence is integrated into machines through artificial intelligence, which is extended to current technologies such as deep learning. Using the concept of AI, machines exhibit intelligent behaviors when processing problems based on prescribed algorithm. There are two types of AI: weak or narrow, and reactive machines, which have limited memories, theories of mind, and self-awareness. In machine learning, machines learn from provided data to make accurate predictions. AI algorithms have been successfully applied to UAVs networks, leveraging their intelligence, to solve challenges related to drones [49]. A recent study [50] discusses the potential integration of AI into UAVs and describes AI-based UAV applications, including supervised, unsupervised, and reinforcement learning schemes. Depending on the controller's interest, a UAV equipped with a camera captures photographs by flying at a lower altitude, allowing control over image resolution.

ML algorithms, however, are required to interpret high-resolution images [51]. ML-based UAVs are discussed in [52] in terms of their importance, scope, and future prospects. In [53], different AI techniques are discussed, including expert systems, ML, distributed AI, and automatic planning-based UAVs. ML has been applied for the intelligent cooperation of UAV swarms in [54] to address their complex and cohesive characteristics. The system begins by developing a digital twin (DT)-based model to predict the physical entity (i.e., the UAV swarm) with high detail and monitor its life-cycle. Subsequently, a decision model based on machine learning is constructed to explore the globally optimal solution and control the behavior of the UAV swarm. This combination presents various challenges associated with the high complexity and heterogeneity of these networks. Additionally, [55] proposes supervised and reinforcement learning strategies for the design of Ultra-Reliable and Low-Latency Communication Radio Access Networks (U-RANs). It explores wireless connectivity and security challenges for various UAV use cases, including delivery systems, real-time multimedia streaming, and UAV-integrated intelligent transportation systems. By using artificial neural networks (ANNs), it enables secure real-time operation with adaptive functionality. With the proposed solutions, UAVs are able to predict future network changes, thereby optimizing their actions to manage their resources efficiently and safely.

10.3.3 Blockchain

For next generation wireless networks to be secure and trusted, blockchain technology is essential. In addition to being adaptable, scalable, immutability, transparent, fast, efficient, and secure, blockchain provides numerous benefits to UAV networks [56]. By using global coverage satellite (GCS) satellites, B5G-enabled UAV-UAV communication is protected from adversarial attacks because shared ledgers are unalterable. The accuracy of stored information is assured through the use of digital distributed ledgers (DDLs), and one-to-one and broadcast communication is protected. Furthermore, blockchains provide chronology, consensus, and audibility, facilitating control, coordination, integrity, and trust among UAV swarms and enabling the exchange of cryptographically cached and secured data between air and ground sensor networks [57]. In case of emergency, a low battery conditions, system fault, or sensor malfunction, UAVs requests for assistance from other UAVs via the blockchain. Additionally, blockchain enables the storage of computations during synchronization. A UAV can download ledger information in an offline manner to optimize processing time and power management capabilities.

10.3.4 Wireless Sensor Networks

Wireless sensor networks (WSNs) are groups of distributed connected sensors that monitor different physical events, such as movement, wind, water flow, temperature, sound, humidity, and more. In smart cities, WSNs monitor pollution levels, traffic, occupancy levels of buildings, energy consumption, and water consumption, among other resources and situations [58]. WSNs play a crucial role in UAVs deployed for various applications in smart cities. UAVs equipped with WSNs enable the collection of real-time data from the environment, which is transmitted wirelessly to ground stations or other network nodes. This integration allows for efficient monitoring and management of smart city infrastructure. Overall, the utilization of WSNs in UAVs for smart cities empowers real-time data collection, remote sensing, and monitoring capabilities. By leveraging the potential of these technologies, smart cities can make informed decisions, implement efficient resource management strategies, and create sustainable urban environments for their residents.

10.3.5 Big Data

In smart cities, decision-making is based on big data, a large amount of information collected from various sources. The concept of big data implies attributes such as a large amount of data, a range of data types, data velocity, and significant data value. A drone is an airborne representation of big data that is used for data generation, data transfer, and data analytics. Big data is gathered from UAS, manned aircraft, satellites, and ground laboratories equipped with new generation sensors during earth-observation. Over the past few years, urban data platforms have played a crucial role in the modern-day governance of cities. The utilization of big data in UAVs for smart cities empowers urban planners, decision-makers, and authorities with valuable insights and information. By harnessing the power of big data analytics, UAVs contribute to improved urban planning, efficient resource management, enhanced public safety, and sustainable development within smart cities.

10.4 Applications of UAV-Based Sustainable Smart Cities

In the past few years, IoT has transformed legacy cities into smart ones, but now it is time for IoD. Drones will power smart cities in the future. The use of drones in smart cities will make them more secure and safe through crime detection. Using flying drones is more reliable and safer than stationary cameras because they can capture the full view and details. By employing drones to fly above the city, every corner will be covered, making surveillance much easier. In addition, drones are capable to monitor traffic on the streets and report accidents. It is possible to use drones to read electric meters, water meters, and gas meters without involving humans. In addition to monitoring air quality, it is employed to measure pollution levels. There are ways to check whether unauthorized construction, gatherings, or other activities are taking place [59]. Applications like smart healthcare, security and privacy, traffic and crowd monitoring, order shipment, infrastructure and environment survey, and disaster management help make cities smarter, as depicted in Fig. 10.4.

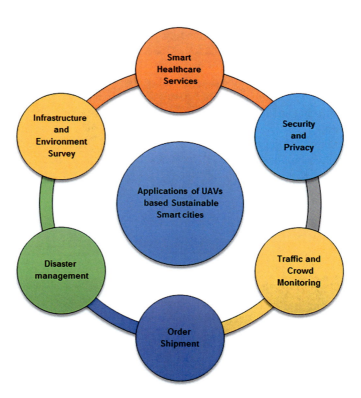

Fig. 10.4 Application of UAV-based sustainable smart cities

10.4 Applications of UAV-Based Sustainable Smart Cities

10.4.1 Smart Healthcare Services

It is essential for UAVs to pick up and deliver vaccines, medicines, and blood samples wherever they are needed. UAVs are currently used to monitor and avoid human interaction with COVID-19 [60, 61]. Furthermore, they enable efficient delivery of healthcare services to patients, including blood, birth control pills, medical supplies to rural areas, vaccines, and snakebite serum [62]. Getting to the hospital on time could make the difference between life and death. UAVs have the ability to reach victims requiring immediate medical attention within minutes. Blood is transported between hospital buildings, and medicine is transported within hospitals using UAVs. Additionally, UAVs offer the healthcare industry exciting possibilities for saving time, money, and lives by providing elderly patients with tools to support them. Medical supplies and food aid are delivered by UAVs to disaster-stricken regions. Their ability to work together with IoT sensors, drones, wearables, and implantable will revolutionize healthcare by continuously monitoring people's health. It is possible to get a diagnosis without traveling miles by utilizing these combined technologies.

10.4.2 Security and Privacy

As smart cities become more complex, managing their security systems and operations has become more difficult. For security monitoring and support, UAVs offers advantages such as more accurate and real-time information that enables authorities to take appropriate actions. In smart cities, UAVs helps police enforce safety and security measures. Their rapid deployment supports these operations in a wide range of situations by collecting real-time information. A UAV is used to assist police in arresting a car thief suspect [63].

10.4.3 Traffic and Crowd Monitoring

Many large cities face traffic congestion as a major problem. Traffic congestion occurs when the number of vehicles rises periodically or suddenly, such as during rush hours, during large events, construction work, or accidents. These problems happens at anytime and anywhere within the cities. Large cities often face traffic congestion as a major problem. During rush hours, large events, construction work, or accidents, the number of vehicles can suddenly increase, leading to this problem. There is no time and place limit to when these problems can occur within these cities. To address these issues, UAVs are utilized to collect and deliver real-time traffic congestion information. The high mobility and visibility of UAVs allow them to provide live feeds covering both the congestion area and their surrounding. The development of traffic simulation models that utilize collected traffic data to evaluate different congestion solutions is currently underway [64]. Integrating this application with other automated systems enable drivers to find or reserve unoccupied parking spots more quickly. In this situation, UAVs can contribute greatly to traffic management. Furthermore, incorporating additional data sources, monitoring techniques, and data collection methods will lead to even better and more accurate results. To enhance the safety of smart cities, UAVs are integrated with various technologies, including video streaming, secure and reliable wireless communication, forensic mapping software, video-based abnormal motion detection [65], and video-based abnormal human behavior recognition. A UAV may be used during large events like parades, sports games, and major outdoor celebrations to monitor crowds. In such events, real-time crowd movement monitoring is vital to security enforcement and safety concerns.

10.4.4 Order Shipment

The delivery of customer orders become more efficient with the use of UAVs. Specifically, designed for this purpose, delivery UAVs enable customers to receive their orders quickly after placing them online. It is easy to transport UAVs to different areas in cities for rapid delivery even in crowded areas. A major benefit of air travel is the ability to avoid congestion areas and take faster routes, which increases mobility. This concept has been developed and tested by companies like Amazon and DHL. Further, Google is developing a system to deliver packages autonomously to different locations within a short period of time called Project Wing. In Queensland, Australia, Google tested rural deliveries [66]. However, there may be security and air traffic control considerations associated with implementation of such applications.

10.4.5 Infrastructure and Environment Survey

City infrastructures are inspected in a flexible, fast, and cost-effective manner using UAVs [67]. Routine inspection of certain city infrastructures are essential in preventing serious problems that may lead to human deaths or incur high repair costs. There are, however, usually a great deal of costs and effort involved in these routine processes. Some of these inspections are conducted with the help of UAVs. Additionally, UAVs are sent to investigate reported problems within these infrastructures at any time. UAVs are utilized to capture images and scan areas that might be problematic. The images and scans are analyzed by software to find indications of infrastructure problems. UAVs were proposed for inspecting bridges, power lines, pipelines, and large buildings [68]. Environmental conditions such as CO_2 levels and air pollution are monitored and measured using UAVs. The risks are accurately assessed, and early detection of harmful conditions is possible.

10.4.6 Disaster Management

A large-scale disaster such as an earthquake, flood, volcano eruption, fire in forests or large infrastructures, or terrorist attack poses many challenges for emergency response management. A disaster situation needs to be monitored, analyzed, arranged for emergency response, and controlled. In most of these cases, humans and emergency teams are unable to reach disaster areas or each other easily and quickly. Meanwhile, many infrastructures are notoriously difficult to maintain, such as emergency response and regular monitoring. These circumstances lend themselves to the use of UAVs [69]. To monitor the current situation and provide real-time information, they are utilized as flexible, reliable, and safe tools. To get better analysis and assessment of the situation, UAVs are deployed at different locations to get more real-time information. It is also possible to use UAVs during some disasters to control the situation. In addition, UAVs carry medical supplies and equipment, locate survivors, and transport injured people as ambulances.

10.5 Open Research Challenges and Research Directions

There are several open issues and challenges associated with UAVs in smart cities, despite the tremendous advantages they offer for delivering services in real time and providing high-reliability services in several applications, reducing energy consumption, reducing delays, and providing efficient bandwidth in several applications. While UAVs offer numerous benefits for smart cities, there are several open research challenges that need to be addressed. Some of these challenges includes: regulations and policies, security and privacy, air traffic management, communication and connectivity, energy efficiency, battery life, autonomous navigation, collision avoidance, scalability, environmental considerations and integration with existing infrastructure. These research challenges require interdisciplinary collaboration and innovation to harness the full potential of UAVs in building smarter and more sustainable cities. Different ways in which UAVs positively impacts a city's services, infrastructures, and residents are achieved through their use in smart cities. UAVs enhance the quality and performance of some smart city services, improving the quality, productivity, timeliness, reliability, and performance of services such as traffic monitoring, merchandise delivery, health emergency services, and wireless communication aided by UAVs. The use of UAVs also improves operations and reduce costs in areas such as remote health emergency support, agriculture management, and environmental monitoring. It is also possible in smart cities to enhance safety and security through the use of UAVs and to save human lives. The utilization of drones for applications such as security and crowd monitoring, health emergency services, and large-scale disaster management makes smart cities a much safer place to live and visit.

Research in the future will examine the potential for extending the use of UAV technology to remote areas, not just cities. Using disruptive technology enables sustainable development of regions with the use of experiences gained in urban environments. In the future, UAVs will be used more widely throughout urban and rural areas, especially in the fields of renewable energy production, industry, transportation, and agriculture, as well as a pseudo-satellite. Recent development efforts are being undertaken to deploy UAVs as air taxis in smart cities, bypassing traffic congestion. With the advancement of technology, all the challenges facing this type of UAV/UAS have the potential to be overcome in the near future. This would make their use in smart cities and regions more relevant.

10.6 Summary

The world and the economy will be transformed by many technologies in the coming decade. There are many technologies that contribute to the development of cities. Smart cities will become better places to live and to do business when UAVs are integrated into them. There are already plans to develop an intelligent government, and with smart cities gaining traction and acceptance, people can expect significant improvements in their day to day lives. As UAV systems are developed and deployed in smart cities, there are still several open issues that need to be addressed. These include wireless sensors, data communications, application management, resource allocation, and power management. We can safely conclude that, when used effectively and efficiently, UAV systems and smart cities have a significant impact and benefit for any country. Using such a solution, we believe that wireless networks and communication will be revolutionized, enabling next-generation, intelligent networks to achieve sustainable smart cities.

References

1. Unmanned Aerial Vehicle (UAV) Market Global Forecast to 2026. [Online]. Available https://www.marketsandmarkets.com/Market-Reports/unmanned-aerialvehiclesuav-market-662.html. Accessed on 8 Mar 2023
2. U.E. Franke, The five most common media misrepresentations of UAVs, in *Hitting the Target*, (2013), pp. 2–13
3. A. Ollero, S. Lacroix, L. Merino, J. Gancet, J. Wiklund, V. Remus, I.V. Perez, L.G. Gutierrez, D.X. Viegas, M.A.G. Benitez, A. Mallet, Multiple eyes in the skies: Architecture and perception issues in the COMETS unmanned air vehicles project. IEEE Robot. Autom. Mag. **12**(2), 46–57 (2005)
4. G.M. Saggiani, B. Teodorani, Rotary wing UAV potential applications: An analytical study through a matrix method. Aircr. Eng. Aerosp. Technol. **76**(1), 6–14 (2004)
5. V. Kharchenko, D. Prusov, Analysis of unmanned aircraft systems application in the civil field. Transport **27**(3), 335–343 (2012)
6. I. Colomina, P. Molina, Unmanned aerial systems for photogrammetry and remote sensing: A review. ISPRS J. Photogramm. Remote Sens. **92**, 79–97 (2014)

7. R.M. Mateos, J.M. Azanon, F.J. Roldan, D. Notti, V. Perez-Pena, J.P. Galve, J.L. Perez-Garcia, C.M. Colomo, J.M. Gomez-Lopez, O. Montserrat, N. Devantery, F. Lamas-Fernandez, F. Fernandez-Chacon, The combined use of PSInSAR and UAV photogrammetry techniques for the analysis of the kinematics of a coastal landslide affecting an urban area (SE Spain). Landslides **14**(2), 743–754 (2017)
8. L. Deng, Z. Mao, X. Li, Z. Hu, F. Duan, Y. Yan, UAV-based multispectral remote sensing for precision agriculture: A comparison between different cameras. ISPRS J. Photogramm. Remote Sens. **146**, 124–136 (2018)
9. P.L. Raeva, J. Sedina, A. Dlesk, Monitoring of crop fields using multispectral and thermal imagery from UAV. Eur. J. Remote Sens. **52**(1), 192–201 (2019)
10. S.C. Hassler, F. Baysal-Gurel, Unmanned aircraft system (UAS) technology and applications in agriculture. Agronomy **9**(10), 618/1–618/21 (2019)
11. M. Gheisari, B. Esmaeili, Applications and requirements of unmanned aerial systems (UASs) for construction safety. Saf. Sci. **118**, 230–240 (2019)
12. S. Bang, H. Kim, UAV-based automatic generation of high-resolution panorama at a construction site with a focus on preprocessing for image stitching. Autom. Constr. **84**, 70–80 (2017)
13. R.C. Erenoglu, O. Erenoglu, N. Arslan, Accuracy assessment of low cost UAV based city modelling for urban planning. Tehnicki Vjesnik—Tech. Gazette **6**, 1708–1714 (2018)
14. C. Warembourg, M. Berger-Gonzalez, D. Alvarez, F. Maximiano Sousa, A. Lopez Hernandez, P. Roquel, J. Eyerman, M. Benner, S. Durr, Estimation of free-roaming domestic dog population size: Investigation of three methods including an Unmanned Aerial Vehicle (UAV) based approach. PLoS One **15**(4), 1–24 (2020)
15. R.K.R. Kummitha, Smart technologies for fighting pandemics: The techno- and human- driven approaches in controlling the virus transmission. Gov. Inf. Q. **37**(3), 101481/1–101481/11 (2020)
16. M. Javaid, A. Haleem, R. Vaishya, S. Bahl, R. Suman, A. Vaish, Industry 4.0 technologies and their applications in fighting COVID-19 pandemic. Diabetes Metab. Syndr. Clin. Res. Rev. **14**(4), 419–422 (2020)
17. I.A.T. Hashem, V. Chang, N.B. Anuar, K. Adewole, I. Yaqoob, A. Gani, E. Ahmed, H. Chiroma, The role of big data in smart city. Int. J. Inf. Manag. **36**(5), 748–758 (2016)
18. A. Ellenberg, A. Kontsos, F. Moon, I. Bartoli, Bridge deck delamination identification from unmanned aerial vehicle infrared imagery. Autom. Constr. **72**, 155–165 (2016)
19. T. Omar, M.L. Nehdi, Remote sensing of concrete bridge decks using unmanned aerial vehicle infrared thermography. Autom. Constr. **83**, 360–371 (2017)
20. J. Vodak, D. Sulyova, M. Kubina, A. Taeihagh, M. De Jong, Advanced technologies and their use in smart city management. Sustainability **13**(10), 5746/1–5746/20 (2021)
21. T. Niedzielski, M. Szymanowski, B. Mizinski, W. Spallek, M. Witek-Kasprzak, J. Slopek, M. Kasprzak, M. Błas, M. Sobik, K. Jancewicz, D. Borowicz, J. Remisz, P. Oettershagen, T. Stastny, T. Hinzmann, K. Rudin, T. Mantel, A. Melzer, B. Wawrzacz, G. Hitz, R. Siegwart, Robotic technologies for solar-powered UAVs: Fully autonomous updraft-aware aerial sensing for multiday search-and-rescue missions. J. Field Robot. **35**(4), 612–640 (2018)
22. A. DeMario, P. Lopez, E. Plewka, R. Wix, H. Xia, E. Zamora, D. Gessler, A.P. Yalin, Water plume temperature measurements by an unmanned aerial system (UAS). Sensors **17**(2), 306/1–306/10 (2017)
23. R. Jurevicius, N. Goranin, J. Janulevicius, J. Nugaras, I. Suzdalev, A. Lapusinskij, Method for real time face recognition application in unmanned aerial vehicles. Aviation **23**(2), 65–70 (2019)
24. L. Silva, O.de, R.A. Bandeira, M. de, V.B.G. Campos, Proposal to planning facility location using UAV and geographic information systems in a post-disaster scenario. Int. J. Disaster Risk Reduct. **36**, 101080/1–101080/11 (2019)
25. S. Mavroulis, E. Andreadakis, N.I. Spyrou, V. Antoniou, E. Skourtsos, P. Papadimitriou, I. Kasssaras, G. Kaviris, G.A. Tselentis, N. Voulgaris, P. Carydis, E. Lekkas, UAV and GIS based rapid earthquake-induced building damage assessment and methodology for EMS-98

isoseismal map drawing: The June 12, 2017 Mw 6.3 Lesvos (Northeastern Aegean, Greece) earthquake. Int. J. Disaster Risk Reduct. **37**, 101169/1–101169/20 (2019)
26. S. Verykokou, C. Ioannidis, G. Athanasiou, N. Doulamis, A. Amditis, 3D reconstruction of disaster scenes for urban search and rescue. Multimed. Tools Appl. **77**(8), 9691–9717 (2018)
27. M. Balasingam, Drones in medicine—The rise of the machines. Int. J. Clin. Pract. **71**(9), 1–4 (2017)
28. R. Madurai Elavarasan, R. Pugazhendhi, Restructured society and environment: A review on potential technological strategies to control the COVID-19 pandemic. Sci. Total Environ. **725**, 138858/1–138858/18 (2020)
29. N. Gaitani, I. Burud, T. Thiis, M. Santamouris, High-resolution spectral mapping of urban thermal properties with unmanned aerial vehicles. Build. Environ. **121**, 215–224 (2017)
30. R. Clarke, Understanding the drone epidemic. Comput. Law Secur. Rev. **30**(3), 230–246 (2014)
31. A. Choi-Fitzpatrick, Drones for good: Technological innovations, social movements, and the state. J. Int. Aff. **68**(1), 19 (2014)
32. C.L. Kang, Y. Cheng, F. Wang, M.M. Zong, J. Luo, J.Y. Lei, The application of UAV oblique photogrammetry in smart tourism: A case study of Longji terraced scenic spot in Guangxi Province. Int. Arch. Photogramm. Remote. Sens. Spat. Inf. Sci. **42**, 575–580 (2020)
33. G. Borkowski, A. Młynarczyk, Remote sensing using unmanned aerial vehicles for tourist-recreation lake evaluation and development. Quaest. Geogr. **38**(1), 5–14 (2019)
34. I. Mademlis, N. Nikolaidis, A. Tefas, I. Pitas, T. Wagner, A. Messina, Autonomous UAV cinematography: A tutorial and a formalized shot-type taxonomy. ACM Comput. Surv. **52**(5), 1–33 (2019)
35. I. Karakostas, I. Mademlis, N. Nikolaidis, I. Pitas, Shot type constraints in UAV cinematography for autonomous target tracking. Inf. Sci. **506**, 273–294 (2020)
36. M. Giurato, P. Gattazzo, M. Lovera, UAV lab: A multidisciplinary UAV design course. IFAC PapersOnLine **52**(12), 490–495 (2019)
37. J.-H. Hong, H.-S. Shin, A. Tsourdos, A design of a short course with COTS UAV system for higher education students. IFAC PapersOnLine **52**(12), 466–471 (2019)
38. K. Park, K. Christensen, D. Lee, Unmanned aerial vehicles (UAVs) in behavior mapping: A case study of neighborhood parks. Urban For. Urban Green. **52**, 126693/1–126693/13 (2020)
39. M.A. Khan, W. Ectors, T. Bellemans, Y. Ruichek, A.-H. Yasar, D. Janssens, G. Wets, Unmanned aerial vehicle-based traffic analysis: A case study to analyze traffic streams at urban roundabouts. Procedia Comput. Sci. **130**, 636–643 (2018)
40. E. Barmpounakis, N. Geroliminis, On the new era of urban traffic monitoring with massive drone data: The pNEUMA large-scale field experiment. Transp. Res. Part C Emerg. Technol. **111**, 50–71 (2020)
41. R. Kamnik, M. Nekrep Perc, D. Topolsek, Using the scanners and drone for comparison of point cloud accuracy at traffic accident analysis. Accid. Anal. Prev. **135**, 105391/1–105391/9 (2020)
42. W.-C. Chiang, Y. Li, J. Shang, T.L. Urban, Impact of drone delivery on sustainability and cost: Realizing the UAV potential through vehicle routing optimization. Appl. Energy **242**, 1164–1175 (2019)
43. H.Y. Jeong, B.D. Song, S. Lee, Truck-drone hybrid delivery routing: Payload-energy dependency and No-Fly zones. Int. J. Prod. Econ. **214**, 220–233 (2019)
44. H.A.E. Al-Jameel, R.R. Muzhar, Characteristics of on-street parking on-street parking in Al-Najaf city urban streets. Transp. Res. Procedia **45**, 612–620 (2020)
45. N.H. Motlagh, T. Taleb, O. Arouk, Low-altitude unmanned aerial vehicles-based internet of things services: Comprehensive survey and future perspectives. IEEE Internet Things J. **3**(6), 899–922 (2016)
46. H.A. Mendoza, A. Ramirez, G.C. Briones, Internet of remote things: A communication scheme for air-to-ground information dissemination, in *Proceedings of IEEE 23rd International Conference on Digital Signal Processing (DSP)*, (Shanghai, China, November 2018), pp. 1–5

References

47. A. Cavoukian, Privacy and drones: Unmanned aerial vehicles, in *Information and Privacy Commissioner of Ontario, Ontario, Canada*, (2012), pp. 1–30
48. W. Crowe, K.D. Davis, A.L. Cour-Harbo, T. Vihma, S. Lesenkov, R. Eppi, E.C. Weatherhead, P. Liu, M. Raustein, M. Abrahamsson, K.S. Johansen, D. Marshal, Enabling science use of unmanned aircraft systems for Arctic environmental monitoring, in *Arctic Monitoring and Assessment Programme*, (2012), pp. 1–30
49. C. Kumar, Artificial intelligence: Definition, types, examples, technologies, Medium.com, (2018). Accessed 12 June 2020. [Online]. Available https://medium.com/@chethankumargn/arti_cialintelligencede_nition-types-examples-technologies-962ea75c7b9b. Accessed on 8 Mar 2023
50. M.-A. Lahmeri, M.A. Kishk, M.-S. Alouini, Artificial intelligence for UAV-enabled wireless networks: A survey. IEEE Open J. Commun. Soc. **2**, 1015–1040 (2021)
51. N. Rey, Combining UAV-imagery and machine learning for wildlife conservation, (M.S. thesis, Ecole Polytechnique Federale de Lausanne, June 2016)
52. A.I. Khan, Y. Al-Mulla, Unmanned aerial vehicle in the machine learning environment. Procedia Comput. Sci. **160**, 46–53 (2019)
53. R. Yin, W. Li, Z.-Q. Wang, X.-X. Xu, The application of artificial intelligence technology in UAV, in *Proceedings of 5th International Conference on Information Science, Computer Technology and Transportation (ISCTT)*, (November 2020), pp. 238–241
54. L. Lei, G. Shen, L. Zhang, Z. Li, Toward intelligent cooperation of UAV swarms: When machine learning meets digital twin. IEEE Netw. **35**(1), 386–392 (2021)
55. V. Kouhdaragh, F. Verde, G. Gelli, J. Abouei, On the application of machine learning to the design of UAV-based 5G radio access networks. Electronics **9**(4), 689/1–689/20 (2020)
56. S. Tanwar, Q. Bhatia, P. Patel, A. Kumari, P.K. Singh, W.-C. Hong, Machine learning adoption in blockchain-based smart applications: The challenges, and a way forward. IEEE Access **8**, 474–488 (2019)
57. S.B. Patel, H.A. Kheruwala, M. Alazab, N. Patel, R. Damani, P. Bhattacharya, S. Tanwar, N. Kumar, BioUAV: Blockchain envisioned framework for digital identification to secure access in next-generation UAVs, in *Proceedings of 2nd ACM MobiCom Workshop Drone Assist. Wireless Commun. 5G Beyond*, (New York, NY, USA, September 2020), pp. 43–48
58. T. Watteyne, K.S. Pister, Smarter cities through standards-based wireless sensor networks. IBM J. Res. Dev. **55**(1), 1–7 (2011)
59. J. Won, S.-H. Seo, E. Bertino, Certificateless cryptographic protocols for efficient drone-based smart city applications. IEEE Access **5**, 3721–3749 (2017)
60. S.H. Alsamhi, B. Lee, Blockchain for multi-robot collaboration to combat COVID-19 and future pandemics. IEEE Access **9**, 44173–44197 (2020)
61. S.H. Alsamhi, B. Lee, M. Guizani, N. Kumar, Y. Qiao, X. Liu, Blockchain for decentralized multi-drone to combat COVID-19 and future pandemics: Framework and proposed solutions. Trans. Emerg. Telecommun. Technol. **32**(9), e4255/1–e4255/19 (2021)
62. T. Mesar, A. Lessig, D.R. King, Use of drone technology for delivery of medical supplies during prolonged field care. J. Spec. Oper. Med. **18**(4), 34–35 (2018)
63. L. Hull, Drone makes first UK 'arrest' as police catch car thief hiding under bushes, *12 Daily Mail*, (2010). Available at http://www.dailymail.co.uk/news/article-1250177/Poli ce-make-arrest-using-unmanned-drone.html. Accessed on 8 Mar 2023
64. B. Coifman, M. McCord, R.G. Mishalani, M. Iswalt, Y. Ji, Roadway traffic monitoring from an unmanned aerial vehicle. IEEE Proc. Intell. Transp. Syst. (IET Digital Library) **153**(1), 11–20 (2006)
65. N. Kiryati, T.R. Raviv, Y. Ivanchenko, S. Rochel, Real-time abnormal motion detection in surveillance video, in *Proceedings of Pattern Recognition, 2008, ICPR 2008, 19th International Conference on IEEE*, (2008), pp. 1–4
66. R. Nieva, S. Rosenblatt, Google spreads its wings, moving into drone deliveries, (2014). Online at http://www.cnet.com/news/google-announces-project-wing-for-drone deliveries/. Accessed on 8 Mar 2023

67. R.L. Mota, L.F. Felizardo, E.H. Shiguemori, A.C. Ramos, F. Mora-Camino, Expanding small uav capabilities with ann: A case study for urban areas inspection. Br. J. Appl. Sci. Technol. **4**(2), 387 (2014)
68. C. Deng, S. Wang, Z. Huang, Z. Tan, J. Liu, Unmanned aerial vehicles for power line inspection: A cooperative way in platforms and communications. Aust. J. Commun. **9**(9), 687–692 (2014)
69. A. Ataei, I.C. Paschalidis, Quadrotor deployment for emergency response in smart cities: A robust MPC approach, in *Proceedings of 2015 IEEE 54th Annual Conference*, (IEEE, 2015), pp. 5130–5135

Index

A
Application layer, 45, 79, 107, 160, 183
Artificial intelligence (AI), vi–x, 3, 14, 17, 30–32, 40, 42, 53, 62, 64, 66–67, 78, 84, 98, 101, 102, 104, 105, 108–110, 112, 113, 119–138, 144, 151, 161, 162, 182, 183, 196, 199, 200, 205, 206, 210, 215, 228

B
Beyond 5G and 6G, 97, 105
Big data, vi, viii, x, 1, 6, 13, 30–32, 41, 42, 45–47, 60–62, 64, 67, 75, 82, 85, 86, 88–90, 92, 104, 148, 151, 161–164, 167, 176, 189, 196, 197, 199, 200, 207–208, 211, 215, 225, 229
Big data analytics, viii, 31, 43, 61, 62, 64, 67, 75, 76, 78, 88–90, 148, 167, 199, 211, 215
Blockchain technology, viii, 3, 13, 30, 31, 42, 47, 53, 61, 62, 64, 68, 101, 102, 104, 114, 136, 161, 165, 197, 208–209, 228, 229
Business layer, 107–108, 183

C
Cloud, 7, 26, 27, 43, 45–47, 59, 65, 66, 81–83, 92, 93, 104, 107, 137, 151, 161–164, 183, 206–207, 210, 215
Cloud computing, viii, x
Cloud computing technology, viii, 16, 21, 30–32, 43, 45, 60, 61, 67, 75, 78, 81–83, 90, 93, 104, 128, 148, 153, 154, 162, 163, 196, 200, 205–207, 234
Cognitive frameworks, 148, 153, 154
Communication layer, 105–106
Communication model, x, 32, 177, 184–189
Communication technologies, v, ix, 31, 53, 64–65, 88, 97, 101, 105, 106, 144, 148, 159, 160, 208, 224
Computing technologies, viii, 30, 31, 65, 81–84, 161, 163
Connected and autonomous, 177, 180–183
Cyber physical systems (CPS), viii, 31, 43, 45, 46, 49, 68

D
Data management layer, 80
Data processing layer, 107
Deep learning (DL), viii, 31, 45, 64, 66–67, 90, 108, 114, 121–124, 126, 127, 133, 136, 164, 228
Disaster management, 110, 225, 230, 233

E
Ecological sustainability, 22–24
Economic sustainability, 20–22
Edge computing, viii, 30, 31, 59, 61, 78, 81–83, 93, 104, 107, 112, 126, 163, 206–207
Energy 4.0, 150, 151
Energy harvesting, ix, 32, 114, 148, 154–156, 168

Energy management, vi, ix, 28, 32, 66, 85, 87, 125, 127, 143, 145, 147, 148, 153, 159, 161, 165, 167, 168
Energy optimization, 148, 155
Energy routing, 148, 156
Energy scheduling, 148, 156
Energy sources, 7, 16, 18, 85, 119, 144, 148, 152, 155–157, 159, 165
Extreme connectivity, ix, 31, 102, 108, 109

F
Fifth-generation (5G), viii, 13, 17, 19, 20, 31, 42, 47, 53, 54, 62, 64, 78, 80, 97–99, 101–103, 105, 108–110, 114, 188
Fog computing, viii, 30, 31, 78, 81, 83, 93, 126, 162, 163, 168, 207, 215

H
Healthcare 1.0, 199, 200
Healthcare 2.0, 199
Healthcare 3.0, 199
Healthcare 4.0, 197, 199–200, 207
Healthcare monitoring devices, 202
Healthcare services, 16–17, 195, 196, 201, 205, 210, 213–215, 231

I
Industry 4.0, viii, 13, 17, 30, 40–44, 46, 59, 145, 233
Industry 5.0, 40, 44
Information communication technology (ICT), viii, 1–3, 9–12, 14, 16, 20–22, 24–26, 28, 31, 42, 48, 53, 59, 61, 62, 65, 79, 124, 164, 221, 224
Infrastructure and environment survey, 230, 232
Intelligent transport systems, vii, 179
International Telecommunication Union (ITU), 19, 145, 206
Internet of energy (IoE), 27, 61, 148–150, 154, 155, 167
Internet of everything (IoE), viii, 30, 31, 64–66, 99, 103, 110, 111, 227
Internet of medical things (IoMT), vi, 197, 200–201, 206
Internet of things (IoT), vi, viii–x, 1, 3, 6, 30–32, 41, 52, 60, 62, 64–66, 75–93, 97–114, 121, 136, 143, 145, 148, 157, 160, 161, 166, 175–177, 180, 186, 196, 197, 199, 205–206, 215, 224, 225, 227–228, 234

applications, 17, 19, 42, 47, 61, 78, 84–92, 104
architecture, 78–81, 90, 154, 158
Internet of vehicles (IoV), vi, vii, ix, x, 32, 175–190

K
Key performance indicators (KPIs), 6, 20–25, 101, 104

L
Layer architecture, 183, 190
Level 0, 177, 180–183
Level 1, 181
Level 2, 181, 182
Level 3, 182
Level 4, 182
Level 5, 177, 180–183
Low power device transceivers, 148, 154

M
Machine learning (ML), viii, ix, 6, 31, 45, 64, 66–67, 75, 84, 87, 101–104, 108–109, 121–124, 126, 128, 132, 133, 135, 137, 138, 162, 164, 199, 201, 207, 213, 228

N
National Institute of Sciences and Technology Policy (NISTEP), 48
Network layer, 104, 105, 157, 159
Network marine and underwater communication extended reality, 111
A Network that can sense, 109
New network architectures, ix, 31, 102, 108, 110
New spectrum technologies, 109

O
Order shipment, 230, 232

P
Personalization, 44, 104, 124, 127, 129, 134, 135, 196, 198–200, 204, 206, 210–211, 213, 215
Physical space, 43

Index

Q
Quantum communication, viii, 31, 65, 98, 103, 105, 106, 114
Quantum computing, 65, 70, 81, 84, 93, 107, 114, 164

R
Receiver design, 148, 155
Robotics, 31, 43, 46, 49, 53, 68, 110, 111, 120, 128, 199, 205, 209, 210, 221, 223

S
Satellite terrestrial, 111, 112
Scheduling optimization, 148, 153
Security and privacy, 29, 52, 59, 90, 213, 214, 230, 231
Security and trust, 108, 110, 114
Sensing layer, 79–81, 104
Sixth-generation (6G), vi, viii, ix, 31, 41, 47, 53, 61, 62, 64, 97–114
6G-IoT framework, 97–114
Smart agriculture, v, ix, 6, 14, 18, 32
Smart applications, 103, 105, 107, 183
Smart building and infrastructure, 6, 14
Smart city applications, 3, 66, 105, 107, 108, 145, 154, 176
Smart city/home, v–x, 1–15, 17–20, 22, 25–33, 40–43, 47, 52, 53, 59–62, 64–70, 76, 80, 82–85, 88–90, 93, 97, 98, 102–104, 107, 111–112, 120, 121, 123–125, 129–135, 137, 138, 143–148, 150, 152–155, 163–168, 175, 176, 179, 180, 189, 190, 195, 196, 211, 221–225, 229–234
Smart education, v, vii, ix, 3, 6, 14, 18, 32, 40
Smart energy, v, vii, ix, 6, 14, 16, 28, 32, 39, 40, 85, 148, 151, 160
Smart equipment, 67
Smart governance, v, ix, 3, 6, 7, 9, 14–15, 32, 225
Smart grid, 3, 16–18, 51, 86, 125, 127, 147, 148, 150–152, 160, 161, 163, 188
Smart healthcare, v, vi, ix, x, 3, 6, 14, 16–17, 28, 32, 40, 112, 195–215, 230
Smart healthcare applications policies and standardization, 213–214
Smart healthcare services, 197, 201–205, 211, 230, 231
Smart homes/living, ix, 32
Smart industries, vii, ix, 6, 14, 17, 32, 107, 111, 126

Smart technology, v, 6, 14, 17, 33, 45
Smart transport, ix, 31, 79, 85, 124–126
Smart transportation, v, 3, 6, 14–16, 27, 28, 40, 53, 61, 107, 108, 224
Social sustainability, 24–25
Society 1.0, 41
Society 2.0, 41
Society 3.0, 41
Society 4.0, 41
Society 5.0, viii, 30, 39–54
Surgery, 110, 112, 205
Sustainable development, vii–ix, 6, 7, 11, 14, 30, 31, 39, 46, 50, 64, 143–145, 147, 234
Sustainable Development Goals (SDGs), viii, 18, 25, 30, 39, 42, 43, 45, 46, 49–51, 54, 70, 89, 97
Sustainable smart cities, vi–x, 1–33, 39–54, 59–70, 75–93, 97–114, 119–138, 143–168, 175–190, 195–215, 221–234

T
Traffic and crowd monitoring, 222, 224, 230–232
Transmission layer, 79–80

U
United Nations (UNs), viii, 39, 42, 49–51, 97, 143
Unmanned aerial vehicles (UAVs), vi, vii, x, 32, 33, 112, 221–234
Use cases, 47, 81, 101–103, 109–113, 148, 150, 165, 187, 200, 228

V
Vehicle to everything (V2E), 176, 184, 187–188
Vehicle-to-grid (V2G), 176, 184, 188
Vehicle-to-home (V2H), 176, 184, 187
Vehicle-to-infrastructure (V2I), 176, 179, 184, 186–187
Vehicle-to-pedestrian (V2P), 184, 188–189
Vehicle-to-roadside (V2R), 176, 184–186
Vehicle-to-vehicle (V2V), 112, 176, 179, 184–187

W
Wireless sensor network (WSN), viii, 31, 62, 64, 68–69, 80, 180, 188, 206, 229